LONDON MATHEMATICAL SOCIETY LECTURE NOTE SERIES

Managing Editor: Professor J.W.S. Cassels, Department of Pure Mathematics and Mathematical Statistics, University of Cambridge, 16 Mill Lane, Cambridge CB2 1SB, England

The books in the series listed below are available from booksellers, or, in case of difficulty, from Cambridge University Press.

4 Algebraic topology, J.F. ADAMS
17 Differential germs and catastrophes, Th. BROCKER & L. LANDER
27 Skew field constructions, P.M. COHN
34 Representation theory of Lie groups, M.F. ATIYAH *et al*
36 Homological group theory, C.T.C. WALL (ed)
39 Affine sets and affine groups, D.G. NORTHCOTT
40 Introduction to H_p spaces, P.J. KOOSIS
42 Topics in the theory of group presentations, D.L. JOHNSON
43 Graphs, codes and designs, P.J. CAMERON & J.H. VAN LINT
45 Recursion theory: its generalisations and applications, F.R. DRAKE & S.S. WAINER (eds)
46 p-adic analysis: a short course on recent work, N. KOBLITZ
49 Finite geometries and designs, P. CAMERON, J.W.P. HIRSCHFELD & D.R. HUGHES (eds)
50 Commutator calculus and groups of homotopy classes, H.J. BAUES
51 Synthetic differential geometry, A. KOCK
54 Markov processes and related problems of analysis, E.B. DYNKIN
57 Techniques of geometric topology, R.A. FENN
58 Singularities of smooth functions and maps, J.A. MARTINET
59 Applicable differential geometry, M. CRAMPIN & F.A.E. PIRANI
60 Integrable systems, S.P. NOVIKOV *et al*
62 Economics for mathematicians, J.W.S. CASSELS
65 Several complex variables and complex manifolds I, M.J. FIELD
66 Several complex variables and complex manifolds II, M.J. FIELD
68 Complex algebraic surfaces, A. BEAUVILLE
69 Representation theory, I.M. GELFAND *et al*
74 Symmetric designs: an algebraic approach, E.S. LANDER
76 Spectral theory of linear differential operators and comparison algebras, H.O. CORDES
77 Isolated singular points on complete intersections, E.J.N. LOOIJENGA
78 A primer on Riemann surfaces, A.F. BEARDON
79 Probability, statistics and analysis, J.F.C. KINGMAN & G.E.H. REUTER (eds)
80 Introduction to the representation theory of compact and locally compact groups, A. ROBERT
81 Skew fields, P.K. DRAXL
82 Surveys in combinatorics, E.K. LLOYD (ed)
83 Homogeneous structures on Riemannian manifolds, F. TRICERRI & L. VANHECKE
86 Topological topics, I.M. JAMES (ed)
87 Surveys in set theory, A.R.D. MATHIAS (ed)
88 FPF ring theory, C. FAITH & S. PAGE
89 An F-space sampler, N.J. KALTON, N.T. PECK & J.W. ROBERTS
90 Polytopes and symmetry, S.A. ROBERTSON
91 Classgroups of group rings, M.J. TAYLOR
92 Representation of rings over skew fields, A.H. SCHOFIELD
93 Aspects of topology, I.M. JAMES & E.H. KRONHEIMER (eds)
94 Representations of general linear groups, G.D. JAMES
95 Low-dimensional topology 1982, R.A. FENN (ed)

96 Diophantine equations over function fields, R.C. MASON
97 Varieties of constructive mathematics, D.S. BRIDGES & F. RICHMAN
98 Localization in Noetherian rings, A.V. JATEGAONKAR
99 Methods of differential geometry in algebraic topology, M. KAROUBI & C. LERUSTE
100 Stopping time techniques for analysts and probabilists, L. EGGHE
101 Groups and geometry, ROGER C. LYNDON
103 Surveys in combinatorics 1985, I. ANDERSON (ed)
104 Elliptic structures on 3-manifolds, C.B. THOMAS
105 A local spectral theory for closed operators, I. ERDELYI & WANG SHENGWANG
106 Syzygies, E.G. EVANS & P. GRIFFITH
107 Compactification of Siegel moduli schemes, C-L. CHAI
108 Some topics in graph theory, H.P. YAP
109 Diophantine Analysis, J. LOXTON & A. VAN DER POORTEN (eds)
110 An introduction to surreal numbers, H. GONSHOR
111 Analytical and geometric aspects of hyperbolic space, D.B.A.EPSTEIN (ed)
112 Low-dimensional topology and Kleinian groups, D.B.A. EPSTEIN (ed)
113 Lectures on the asymptotic theory of ideals, D. REES
114 Lectures on Bochner-Riesz means, K.M. DAVIS & Y-C. CHANG
115 An introduction to independence for analysts, H.G. DALES & W.H. WOODIN
116 Representations of algebras, P.J. WEBB (ed)
117 Homotopy theory, E. REES & J.D.S. JONES (eds)
118 Skew linear groups, M. SHIRVANI & B. WEHRFRITZ
119 Triangulated categories in the representation theory of finite-dimensional algebras, D. HAPPEL
121 Proceedings of *Groups - St Andrews 1985*, E. ROBERTSON & C. CAMPBELL (eds)
122 Non-classical continuum mechanics, R.J. KNOPS & A.A. LACEY (eds)
123 Surveys in combinatorics 1987, C. WHITEHEAD (ed)
124 Lie groupoids and Lie algebroids in differential geometry, K. MACKENZIE
125 Commutator theory for congruence modular varieties, R. FREESE & R. MCKENZIE
126 Van der Corput's method for exponential sums, S.W. GRAHAM & G. KOLESNIK
127 New directions in dynamical systems, T.J. BEDFORD & J.W. SWIFT (eds)
128 Descriptive set theory and the structure of sets of uniqueness, A.S. KECHRIS & A. LOUVEAU
130 Model theory and modules, M. PREST
131 Algebraic, extremal & metric combinatorics, M-M. DEZA, P. FRANKL & I.G. ROSENBERG (eds)
132 Whitehead groups of finite groups, ROBERT OLIVER
133 Linear algebraic monoids, MOHAN S. PUTCHA
134 Number theory and dynamical systems, M. DODSON & J. VICKERS (eds)
135 Operator algebras and applications, 1, D. EVANS & M. TAKESAKI (eds)
136 Operator algebras and applications, 2, D. EVANS & M. TAKESAKI (eds)
137 Analysis at Urbana, I, E. BERKSON, T. PECK, & J. UHL (eds)
138 Analysis at Urbana, II, E. BERKSON, T. PECK, & J. UHL (eds)
139 Advances in homotopy theory, S. SALAMON, B. STEER & W. SUTHERLAND (eds)
140 Geometric aspects of Banach spaces, E.M. PEINADOR & A. RODES (eds)
141 Surveys in combinatorics 1989, J. SIEMONS (ed)
142 The geometry of jet bundles, D.J. SAUNDERS
143 The ergodic theory of discrete groups, P.J. NICHOLLS

London Mathematical Society Lecture Note Series. 141

Surveys in Combinatorics, 1989

Edited by

Johannes Siemons
School of Mathematics
University of East Anglia

CAMBRIDGE UNIVERSITY PRESS
Cambridge
New York Port Chester Melbourne Sydney

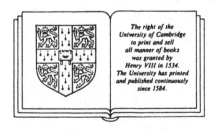

CAMBRIDGE UNIVERSITY PRESS
Cambridge, New York, Melbourne, Madrid, Cape Town,
Singapore, São Paulo, Delhi, Tokyo, Mexico City

Cambridge University Press
The Edinburgh Building, Cambridge CB2 8RU, UK

Published in the United States of America by
Cambridge University Press, New York

www.cambridge.org
Information on this title: www.cambridge.org/9780521378239

First published 1989

A catalogue record for this publication is available from the British Library

Library of Congress Cataloguing in Publication data

ISBN 978-0-521-37823-9 Paperback

PREFACE

Since its beginning in 1969 the British Combinatorial Conference has grown into an established international meeting. This year the twelfth conference is being held in Norwich under the auspices of the School of Mathematics at the University of East Anglia. Participants come from a great number of countries worldwide and represent a multitude of interests in combinatorial theory.

This volume contains the contributions of the principal speakers.They were invited to prepare a survey paper for this book and to deliver a lecture in an area of their expertise. In this way it is hoped to make available a valuable source of reference to the current state of art in combinatorics. The speakers have produced their papers well in advance so that they are now all available in time for the conference.

This book has been produced to a tight schedule. I am grateful to the authors for their cooperation and to the referees for their assistance and comments about the papers. The British Combinatorial Conference is largely self-financing but on behalf of the committee I would like to thank the London Mathematical Society, Norwich Union and Peat Marwick McLintock for their financial support.

Johannes Siemons
Norwich April 1989

CONTENTS

On the theory of designs

E. F. ASSMUS, JR.

INTRODUCTION

Several years ago I was asked a seemingly innocuous question: *What are the minimal-weight vectors of the code of an affine plane?* I thought the answer would be that they were, just as in the projective case, simply the scalar multiples of the lines; indeed, that may be true and the question is still open. I managed to prove this (for arbitrary affine planes) only for those of prime order.

The question is deeper than it at first seems. If, for example, one could prove that the minimal-weight vectors of the code of an arbitrary affine plane were simply the scalar multiples of the lines, one would have a proof of the fact [**15**] that a projective plane of order ten has no ovals; indeed, one would prove that no projective plane of order congruent to two modulo four, except the Fano plane, could have an oval. (The minimal-weight vectors of the code of a *desarguesian* affine plane are the scalar multiples of the lines of the plane but the only known proof relies heavily on algebraic coding theory.)

These considerations led J. D. Key (who asked the original question) and me to what seems to be a fruitful approach to affine and projective planes and to what we hope will be a fruitful approach to the theory of designs. The purpose of this paper is to explain these matters. Much of the work we have done has already appeared and thus the present paper will rely heavily on four joint papers : *Arcs and ovals in hermitian and Ree unitals*, *Affine and projective planes*, *Baer subplanes, ovals and unitals*, and *Translation planes and derivation sets*. A *brief* summary of some of the material contained in these four papers can be found in [**1**].

In the first section the original definition of a symmetric design is given, the congruence constraint explained, and some remarks on the layout of tables of designs made. (These remarks are expanded on in the Appendix.) The second section defines the hull of a design and the third explains the use of this notion in the investigation of affine planes. The Hamada-Sachar Conjecture and translation planes are then discussed and, following that, derivations are put into the current setting. Finally a few concluding remarks are made.

THE CONGRUENCE CONDITION FOR A SYMMETRIC DESIGN

Tables of possible parameter sets of designs, together with a token design when existence is known, appear from time to time. The layout of these tables usually takes a form more suitable for the design of experiments than for theoretical analysis. Although I have no intention of producing yet another table I do want to suggest a different layout that may throw more light on the theoretical aspects of design theory. I do this here only for so-called symmetric designs but the Appendix shows how to carry out a similar approach for more general designs.

A symmetric (v, k, λ)-design, where v, k, and λ are integers with $1 < k < v - 1$, consists of two disjoint v-sets, $\mathcal{P}$ and $\mathcal{B}$, called points and blocks, with a regular, bipartite graph of valency k imposed, $\mathcal{P}$ and $\mathcal{B}$ being the two parts. The graph satisfies the further condition that given any two distinct points there are precisely λ paths of length two from one point to the other. It follows that the same property holds for any two distinct blocks (hence the unfortunate term "symmetric design"). Thus, one has no reason to call the elements of $\mathcal{P}$ points and those of $\mathcal{B}$ blocks for it might as well have been the reverse. Moreover, the complementary bipartite graph (eliminate the given edges (P, B) and introduce, instead, those (Q, C) which weren't originally edges) is also a symmetric design—with parameters

$$(v, v-k, \frac{1}{\lambda}(k-\lambda)(k-\lambda-1)).$$

So symmetric designs come in pairs, a design and its complement, and in fours if one wishes (as I do) to distinguish $(\mathcal{P}, \mathcal{B})$ from $(\mathcal{B}, \mathcal{P})$, its *dual*.

Given a (v, k, λ) design on $(\mathcal{P}, \mathcal{B})$ one can associate to each $B \in \mathcal{B}$ the k-subset of $\mathcal{P}$ given by $\{P \in \mathcal{P} | (P, B) \text{ is an edge}\}$ and these v k-subsets of $\mathcal{P}$ are (again) called blocks. Because of the path condition, any two blocks meet in λ points and any two points are contained in λ blocks. Moreover, the path condition immediately implies that

$$\lambda(v-1) = k(k-1), \qquad *$$

the congruence condition relating the parameters.

The most important parameter of a symmetric design has not yet appeared. It is the *order*, namely $k - \lambda$, of the design and it is denoted by n. Observe that * implies that $n(n-1)$, clearly congruent to $k(k-1)$ modulo λ, is congruent to 0 modulo λ and, writing $n(n-1) = \lambda\mu$, the parameters are

$$(2n + \lambda + \mu, n + \lambda, \lambda),$$

with the parameters of the complementary design being

$$(2n + \lambda + \mu, n + \mu, \mu).$$

This suggests listing the possible parameter sets for symmetric designs via the order n, using the divisors, λ, of $n(n-1)$ in turn, possibly eliminating the complementary divisor. Here's how such a table would begin:

Order	Parameter sets
2	(7, 3, 1), (7, 4, 2)
3	(13, 4, 1), (13, 9, 6)
	(11, 5, 2), (11, 6, 3)
4	(21, 5, 1), (21, 16, 12)
	(16, 6, 2), (16, 10, 6)
	(15, 7, 3), (15, 8, 4)

For each order the parameters of the possible projective plane and its complement appear first and the parameters of the Hadamard designs (i.e., designs with $\lambda = n-1$ or n) last. Moreover, the organization suggested permits one easily to examine the parameter sets for any particular order. Order twelve is interesting. Here, omitting complements, one gets

$$\begin{gathered}(157, 13, 1),\\(92, 14, 2),\\(71, 15, 3),\\(61, 16, 4),\\(52, 18, 6),\\(47, 23, 11).\end{gathered}$$

The Bruck-Ryser-Chowla Theorem rules out $\lambda = 2$ and $\lambda = 6$, but Beker and Haemers [7] have constructed a design with $\lambda = 3$ and van Trung [18] one with $\lambda = 4$. Of course, the Hadamard design exists (presumably, an enormous number) and hence only the plane of order twelve is in doubt.

Were one to list parameter sets by increasing order and sieve with the Bruck-Ryser-Chowla Theorem, then $(157, 13, 1)$ would be only the second question mark, the first being the plane of order ten. Put another way, all symmetric designs that could exist do exist through order twelve, excepting, possibly, the projective planes of orders ten and twelve. It has been recently reported that the computer calculation undertaken by Clement Lam et al has been completed and that the result is that there is no projective plane of order ten; the other two symmetric designs of order ten that could exist, do exist; they are the Hadamard design and the design with parameters (41, 16, 6).

The designs of orders two and three are unique and the precise number of designs is known through order six. Roughly speaking, for a given order, existence gets easier and more designs exist (if they can at all) as λ grows (up to, of course, $n-1$). For order four, for example, the plane is unique, there are three biplanes, and five triplanes. And, more generally, it is conjectured that there are Hadamard designs for every possible order.

THE CODE AND THE HULL OF A DESIGN

To bring algebraic coding theory into play notational compromises must be made. The concern is with arbitrary designs (for a definition see the Appendix): $|\mathcal{P}|$, the cardinality of the point set, will be denoted by N (rather than v) and the cardinality of a block will, generally, be denoted by d (rather than k). The number of blocks through two points— λ— and the difference between the number of blocks through one point and the number through two—the order n of the design—will be denoted in the standard way.

So, given $\mathcal{D} = (\mathcal{P}, \mathcal{B})$, a design of order n, and any field F, let $F^{\mathcal{P}}$ be the vector space of all functions from $\mathcal{P}$ to F with, of course, point-wise addition and scalar multiplication. For $v \in F^{\mathcal{P}}$, denote the value of v at the point P by v_P and, for X a subset of $\mathcal{P}$, denote by v^X the characteristic function of X. Thus $v_P^X = 1$ if $P \in X$ and 0 otherwise.

Now denote by

$$C_F(\mathcal{D})$$

the subspace of $F^{\mathcal{P}}$ generated by all v^B where B is a block of $\mathcal{D}$ and call this subspace the *code of $\mathcal{D}$ over F*. If the dimension of $C_F(\mathcal{D})$ is k, then it is an (N, k)-code in the language of algebraic coding theory, N being the *length* of the code.

Moreover, its minimum weight is at most d. Here, the *weight* of a vector $v \in F^{\mathcal{P}}$, $wgt(v)$, is $|\{P \in \mathcal{P} | v_P \neq 0\}|$ or, in other words, the cardinality of the support of the function v. The *minimum weight* of any subspace C of $F^{\mathcal{P}}$ is

$$\mathrm{Min}\{wgt(c) | c \in C, c \neq 0\}.$$

Equip $F^{\mathcal{P}}$ with the standard inner product: $[v, w] = \sum_{Q \in \mathcal{P}} v_Q w_Q$. For C a subspace of $F^{\mathcal{P}}$, $C^{\perp}$ denotes the subspace orthogonal to C:

$$C^{\perp} = \{v \in F^{\mathcal{P}} \,|\, [v, c] = 0 \text{ for all } c \in C\}.$$

Finally, set

$$Hull_F(\mathcal{D}) = C_F(\mathcal{D}) \cap C_F(\mathcal{D})^\perp$$

and call this subspace the *hull of* $\mathcal{D}$ *over* F.

EXAMPLE: If $\mathcal{D} = AG_2(\mathbf{F}_4)$, the affine plane of order four, then $Hull_{\mathbf{F}_2}(\mathcal{D})$ is the $(16,5)$ extended binary Reed-Muller code and $Hull_F(\mathcal{D}) = \{0\}$ for any field F of characteristic other than two.

To obtain non-trivial hulls one must choose fields with characteristic dividing n [17]. The choice will always be $F = \mathbf{F}_p$, p a prime; $Hull_F(\mathcal{D})$ will be denoted by $Hull_p(\mathcal{D})$ and referred to as the *hull at* p *of* $\mathcal{D}$. Similarly, $C_p(\mathcal{D})$ will replace $C_F(\mathcal{D})$ and the subscript will be omitted if p is clear from the context.

If σ is any permutation of $\mathcal{P}$ (i.e. if $\sigma \in Sym(\mathcal{P})$) then σ acts naturally on $F^{\mathcal{P}}$ via

$$(v\sigma)_P = v_{\sigma(P)}.$$

(Observe that σ acts on $\mathcal{P}$ on the left and on $F^{\mathcal{P}}$ on the right.) Clearly, if σ is an automorphism of $\mathcal{D}$ (which means that $\sigma(B)$ is a block whenever B is) then σ leaves the subspace $C(\mathcal{D})$ invariant. It can very easily happen—and in non-trivial ways—that the subgroup of $Sym(\mathcal{P})$ leaving $C(\mathcal{D})$ invariant is larger than the subgroup leaving $\mathcal{D}$ invariant. Moreover, the subgroup leaving the hull invariant may be still larger.

EXAMPLE: If $\mathcal{H}$ is the unital on 28 points given by the unitary group $U_3(\mathbf{F}_3)$ then the symplectic group $Sp_6(\mathbf{F}_2)$ leaves $C_2(\mathcal{H})$ invariant. If $\mathcal{R}$ is the Ree unital on 28 points the group leaving $C_2(\mathcal{R})$ invariant is the small Ree group $P\Gamma L_3(\mathbf{F}_8)$ while the group leaving the hull invariant is again the symplectic group.

This example was decisive in the genesis of the notion of the hull of a design, for these two unitals have isomorphic hulls despite the fact that they are not isomorphic as designs. See [2] for the details.

One of the reasons that it was not previously observed that the hull of a design is as important as the code of a design was the fact that, *for symmetric designs*, the hull is intimately related to the code. We end this section with a result that substantiates this assertion.

THEOREM 1. *Let $\mathcal{D}$ be a symmetric design and $\mathcal{D}_{comp}$ the complementary design. Let p be a prime dividing n, their common order. Then, if p does not divide d, the block size of $\mathcal{D}$, $Hull_p(\mathcal{D}) = C_p(\mathcal{D}_{comp})$, it is of codimension one in $C_p(\mathcal{D})$, and it consists of those vectors in $C_p(\mathcal{D})$ with $\sum_{P\in\mathcal{P}} v_P = 0$. If p does divide d, then $C_p(\mathcal{D}) \subseteq C_p(\mathcal{D})^\perp$ and $Hull_p(\mathcal{D}) = C_p(\mathcal{D})$.*

PROOF: Let $\mathcal{D}$'s parameters be (N, d, λ). Now $d = \lambda + n$, $N = 2n + \lambda + \mu$

where $n(n-1) = \lambda\mu$ and $\mathcal{D}_{comp}$'s parameters are $(N, n+\mu, \mu)$. Clearly, if p does not divide d, it does not divide λ and, since it must divide $n(n-1)$, it must divide μ. Hence p divides $n+\mu$, the block size of $\mathcal{D}_{comp}$. This shows that $C(\mathcal{D}_{comp}) \subset C(\mathcal{D}_{comp})^{\perp}$. Reversing the rôles of $\mathcal{D}_{comp}$ and $\mathcal{D}$ yields the last assertion of the Theorem. Continuing the argument, we now have, since p does not divide d, that

$$d^{-1} \sum_{B \in \mathcal{B}} v^B = J,$$

where J is the *all-one* vector. It follows, since $J \in C(\mathcal{D})$, that $C(\mathcal{D}_{comp}) \subset C(\mathcal{D})$. Moreover, $C(\mathcal{D}) = C(\mathcal{D}_{comp}) \oplus \mathbf{F}_p J$, the sum being direct because $[J, J] = N$, a non-zero element of $\mathbf{F}_p$. Thus $C(\mathcal{D}_{comp})$ is of codimension one in $C(\mathcal{D})$ and, because $\sum_{P \in \mathcal{P}} v_P = 0$ whenever $v \in C(\mathcal{D}_{comp})$,

$$C(\mathcal{D}_{comp}) = \{v \in C(\mathcal{D}) | \sum_{P \in \mathcal{P}} v_P = 0\} = Hull(\mathcal{D}).$$

REMARKS:

(1) Since $(J - v^B) - (J - v^C) = v^C - v^B$, it follows easily, when p does not divide d, that $Hull(\mathcal{D})$ is generated by the vectors of the form $v^C - v^B$, B and C being blocks of $\mathcal{D}$.

(2) It can happen that p divides both λ and μ and hence all the parameters. (This occurs, for example, for the $(16, 6, 2)$ designs.) In this case it is easy to see that $C(\mathcal{D}) = C(\mathcal{D}_{comp})$ if and only if J is in both codes and then

$$Hull(\mathcal{D}) = C(\mathcal{D}) = C(\mathcal{D}_{comp}) = Hull(\mathcal{D}_{comp}).$$

If J is in neither code then $C(\mathcal{D}) \cap C(\mathcal{D}_{comp})$ is of codimension one in the code of each design. Moreover, this intersection is generated by vectors of the form $v^B - v^C$ where B and C are both blocks of $\mathcal{D}$ (or both blocks of $\mathcal{D}_{comp}$). I do not have an example of this phenomenon. It would be very interesting to know—even for $p = 2$—precisely when J is in the code of the design.For example, were one to show that that J were in the code of a design with parameters $(2^{2m}, 2^{2m-1} - 2^{m-1}, 2^{2m-2} - 2^{m-1})$ and dimension $2m+2$ one would have the following improvement of a result of Dillon and Schatz [**11**, Corollary 1].

A symmetric design has parameters $(2^{2m}, 2^{2m-1} - 2^{m-1}, 2^{2m-2} - 2^{m-1})$ and code of dimension $2m+2$ if and only if it can be obtained from the Reed-Muller code of length 2^{2m} and a difference set in the elementary abelian 2-group of order 2^{2m} as the vectors of minimal weight in the code spanned by the Reed-Muller code and the difference set.

Were this result true it would characterize the designs of minimal 2-rank with the given parameters and it could be viewed as a theorem of Hamada-type (see Theorem 5 below) but one should be aware of the fact [**12**] that the number of such designs goes to infinity with m —in contrast with planes where, conjecturally, the design of minimum rank is unique.

(3) The Theorem gives the relationship between the codes of a symmetric design and its complement. The relationship between the codes of a symmetric design and its dual has not been thoroughly investigated; obviously the dimensions are equal, but, beyond that, only fragmentary results are known at present. An interesting example of what can happen is given by the two triplanes of order four which are duals of one another: the weight distributions of the two codes are identical but they are not isomorphic (see the discussion preceding Theorem 4).

THE HULL OF AN AFFINE PLANE

The importance of the hull in design theory is best illustrated when the design is an affine plane. Given such a plane π it determines a unique projective plane Π and a line of that plane L_∞ called the line at infinity. Both π and Π have the same order, n say, and, as designs, their parameters are

$$(n^2, n, 1) \text{ and } (n^2 + n + 1, n + 1, 1).$$

If the point sets of Π and π are $\mathcal{P}$ and $\mathcal{A}$, respectively, then $\mathcal{A}$ is simply $\mathcal{P} \setminus L_\infty$. There is a natural linear transformation from $F^{\mathcal{P}}$ to $F^{\mathcal{A}}$ given by simply restricting the functions on $\mathcal{P}$ to $\mathcal{A}$. This transformation relates the codes, the hulls, and their orthogonals. Precisely, we have the following result:

PROPOSITION 1. *Let π be an affine plane of order n and Π its projective completion with L_∞ the line at infinity. Then, for a prime p dividing n, both $C_p(\pi)$ and $Hull_p(\pi)^\perp$ are, respectively, the images of $C_p(\Pi)$ and $Hull_p(\Pi)^\perp$ under the natural transformation of $F^{\mathcal{P}}$ onto $F^{\mathcal{A}}$. Further, $Hull_p(\pi)$ is the image of the subcode $\{c \in Hull_p(\Pi) \mid c_Q = 0 \text{ for } Q \text{ on } L_\infty\}$, $\dim Hull_p(\pi) = \dim C_p(\pi) - n$, and $Hull_p(\pi)$ is generated by all $v^\ell - v^m$ where ℓ and m are two parallel lines of π.*

A proof of this result can be found in [**3**]. One important point here is that the orthogonal of the hull of Π has minimum weight $n+1$ and the minimal-weight vectors are *all* of the form αv^L where L is a line of Π and α a non-zero element of $\mathbf{F}_p$, but the orthogonal of the hull of π (although it has minimum weight n, as expected) has minimal-weight vectors that are not necessarily, in fact not usually, of the form αv^{ℓ} where ℓ is a line of π. They *are* of the form αv^{X}, however, where X sometimes has a nice geometric interpretation. Vectors of the form αv^{X} are called *constant* vectors and are referred to, by an abuse of language, as *scalar multiples of* X.

A second important point is the fact that, of the codes involved, the affine hull is the one of smallest dimension and it is generated by the differences of parallel lines and hence easily computable. The nature of its minimal-weight vectors is of crucial importance: for example, were it always true that the minimum weight of the affine hull were $2n$ and that the minimal-weight vectors were precisely the scalar multiples of the above generators, then one would recover the projective plane from the affine hull for p odd and the theory we are sketching would probably be useless. For desarguesian affine (or projective) planes it *is* true that the minimum weight of the hull is $2n$ and that there are no unexpected minimal-weight vectors; moreover, to establish these results one makes heavy use of algebraic coding theory. More generally, the hull of an affine translation plane has minimum weight $2n$ (a fact that follows easily from Proposition 3 below) but it is not, in general, true that there are only the expected minimal-weight vectors; for example, the hull of the non-desarguesian affine translation plane of order nine has unexpected minimal-weight vectors and the plane cannot be recovered from the hull. It is to be noted that an arbitrary affine plane of prime order can be recovered from its hull; perhaps this could be viewed as evidence that such a plane is desarguesian although my own view is that it is too early to speculate. The following result, whose proof can be found in [3], details part of the story:

THEOREM 2. *Let* π *be an affine plane of order* n *and* p *a prime dividing* n. *Then the minimum weight of* $Hull_p(\pi)^{\perp}$ *is* n *and all minimal-weight vectors are constant. Moreover,*

(1) *If* $n = p$, *then the minimal-weight vectors of* $Hull_p(\pi)^{\perp}$ *are precisely the scalar multiples of the lines of* π *and the hull uniquely determines the plane.*
(2) *If* $n = p^2$, *then the minimal-weight vectors of* $Hull_p(\pi)^{\perp}$ *are scalar multiples of either lines or Baer subplanes of* π.

In the desarguesian case the minimal-weight vectors of $C_p(AG_2(\mathbf{F}_q))$, *where* $q = p^s$, *are scalar multiples of the lines of* $AG_2(\mathbf{F}_q)$ *and the minimal-weight vectors of the hull are precisely the scalar multiples of vectors of the form* $v^{\ell} - v^{m}$ *where* ℓ *and* m *are two parallel lines.*

In general one cannot say a great deal about the dimension of the code of an affine plane, but the dimension is known if the plane is desarguesian: if $q = p^s$, p a prime, then the dimension of the code of the desarguesian affine plane of order q is

$$\binom{p+1}{2}^s.$$

A conjecture, due independently to Hamada and Sachar, states that the code of any affine plane of order q has at least this dimension, with equality only if it *is* the desarguesian plane. The next Section will discuss this matter further.

What one can say in general is that whenever p but not p^2 divides the order of the plane, π say, then dim $C_p(\pi)$ is $\frac{1}{2}n(n+1)$ and hence, by Proposition 1 and Theorem 2, the classification of planes of prime order p is tantamount to the classification of certain $[p^2, \frac{1}{2}p(p-1)]$-codes.

THE HAMADA-SACHAR CONJECTURE AND TRANSLATION PLANES

Let π be an affine plane of order n and H its hull at p for some prime p dividing n. Then, as indicated above, $H^\perp$ usually contains many more minimal-weight vectors than simply the scalar multiples of lines. Thus if $n = p^2$, we have the Baer subplanes appearing and, if n is even and $p = 2$, the hyperbolic ovals. (A *hyperbolic oval* of π is a set of $n+2$ points with two at infinity and no three collinear.) Many other configurations arise as well (see [5]).

The collection of supports of these constant vectors (the *support* of a vector v is $\{P \in \mathcal{P} | v_P \neq 0\}$) may very well contain affine planes other than the one with which one started. For example, $Hull_2(AG_2(\mathbf{F}_4))$ is the (16,5) binary Reed-Muller code with weight distribution $1+30X^8+X^{16}$ and $Hull_2(AG_2(\mathbf{F}_4))^\perp$ is the (16,11) binary extended Hamming code with 140 weight-4 vectors. These vectors form a Steiner quadruple system but, as a *two-design*, they contain 112 subdesigns which are affine planes of order four, all of which have the same hull. The 20 vectors forming an affine plane of order four having been chosen, the remaining 120 are Baer subplanes of that plane. Of course, no new affine planes occur in this simple example, but in the next possible case, order nine, interesting things do occur: there are four projective planes of order nine and seven affine planes of order nine; the two translation planes of order nine each have an affine part that yields the other in the way indicated—and the other two projective planes do likewise.

In order to discover and perhaps classify planes of order n one should have at one's disposal linear codes of length $N = n^2$ over $\mathbf{F}_p$ (where p divides n). The minimum weight should be n, and there must be a sufficiently rich structue of minimal-weight vectors to accommodate $C_p(\pi)$'s for various affine planes π. Of course, given an affine plane π, $B = Hull_p(\pi)^\perp$ is such a code. Fortunately, for $n = p^s$ there are very standard choices for such codes B and, more importantly, these codes have been intensively studied by algebraic coding theorists, in particular, by Philippe Delsarte [8] and Delsarte et al [9]. The results contained in these two papers were crucial to the development of the ideas here presented. In particular, without the dimensions of what are here called the *standard choices* no bounds on the ranks of the incidence matrices of translations planes would be available and, without the results on the minimal-weight vectors of these standard choices, the notion of *tame* would not have surfaced and the weak version of Hamada-Sachar would not have been proved. These papers are rather difficult to read—especially for finite geometers; a brief outline of the results one needs is contained in Appendix I of [3].

In order to properly discuss these linear codes and the affine planes connected with them it is best to make a few definitions.

DEFINITION 1. *Let B be an arbitrary code of length n^2 over $\mathbf{F}_p$ (where p divides n) with minimum weight n.*

(1) *An affine plane π of order n is said to be contained in B if $C_p(\pi)$ is code isomorphic to a subcode of B.*
(2) *An affine plane π of order n is said to be linear over B if $C_p(\pi)$ is code isomorphic to a subcode C of B where $B \subseteq C + C^\perp$.*

EXAMPLE: If one takes the Veblen-Wedderburn plane Ψ of order nine and a "real" line R and sets $\psi = \Psi^R$, the affine plane with R its line at infinity, then the vectors in $Hull(\psi)^\perp$ of the form v^S, where S is a line or Baer subplane of ψ, generate an [81,48] ternary code B over which ψ is linear. Moreover, $\omega = \Omega^L_{dual}$ is also linear over B where Ω is the non-desarguesian translation plane of order nine, Ω_{dual} its dual, and L a line through the translation point. Both $Hull(\psi)^\perp$ and $Hull(\omega)^\perp$ are [81,50] ternary codes containing B. They are not code-isomorphic since $Hull(\psi)^\perp$ contains 2×306 minimal-weight vectors while $Hull(\omega)^\perp$ contains 2×522. The 2×306 weight-9 vectors are the scalar multiples of the 90 lines and the 216 Baer subplanes of ψ; that ψ has 216 Baer subplanes follows from an easy counting argument and the facts contained in [10]. A similar situation obtains for ω. By properly choosing subcodes of B isomorphic to the codes of ψ and ω one gets the following Hasse diagram.

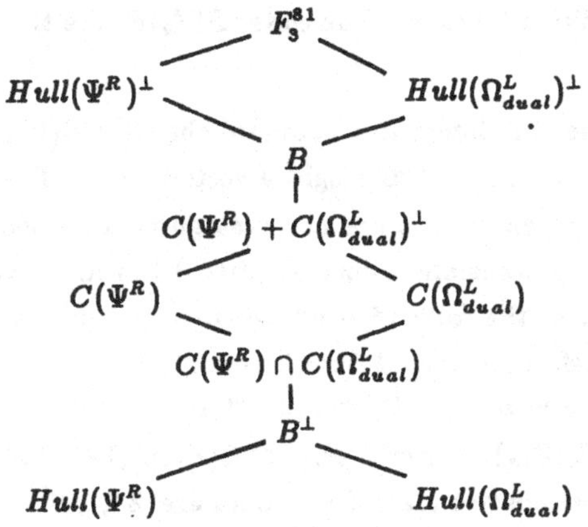

The code B of the above example is not a *standard choice* and we turn now to such codes. Set $q = p^s$ and let F be an arbitrary subfield of $\mathbf{F}_q$. Let V be a 2-dimensional vector space over $\mathbf{F}_q$ and consider V as a vector space over F. It will, of course, have dimension $2[\mathbf{F}_q : F]$, where $[K : F]$ denotes the degree of K over F. Set $m = [\mathbf{F}_q : F]$ and consider the collection, $\mathbf{L}_F$, of all m-dimensional subspaces over F and their translates under the addition in V, i.e., all the m-dimensional cosets in the affine space of V, $AG_{2m}(F)$.

Now $\mathbf{F}_p^V$ is a q^2-dimensional vector space over $\mathbf{F}_p$. Each $X \in \mathbf{L}_F$ is a subset of V of cardinality q and defines a vector v^X of $\mathbf{F}_p^V$. Let $B(\mathbf{F}_q|F)$ be the subspace of $\mathbf{F}_p^V$ spanned by $\{v^X \mid X \in \mathbf{L}_F\}$. $B(\mathbf{F}_q|F)$ is a code of length q^2 over $\mathbf{F}_p$ with minimum weight q, its minimal-weight vectors are scalar multiples of the vectors v^X, and its dimension is computable. Let $E(\mathbf{F}_q|F)$ be the subcode of $B(\mathbf{F}_q|F)$ generated by vectors of the form $v^X - v^Y$ where X and Y are translates of the same m-dimensional subspace of V, viewed as a vector space over F. Once again, the minimal-weight vectors are scalar multiples of these generators and the dimension is computable.

If F and K are two subfields of $\mathbf{F}_q$, $q = p^s$, with $F \subseteq K$, then clearly, $B(\mathbf{F}_q|K) \subseteq B(\mathbf{F}_q|F)$. At one extreme, $F = \mathbf{F}_q$, $\mathbf{L}_F$ consists of the lines of the affine plane $AG_2(\mathbf{F}_q)$ and $B(\mathbf{F}_q|\mathbf{F}_q) = C_p(AG_2(\mathbf{F}_q))$. At the other extreme, $F = \mathbf{F}_p$, $\mathbf{L}_F$ consists of all s-dimensional subspace and their translates, where V is viewed as a $2s$-dimensional vector space over $\mathbf{F}_p$.

The mapping $F \mapsto B(\mathbf{F}_q|F)$ establishes a Galois correspondence between the subfields of $\mathbf{F}_q$ and certain subcodes of $\mathbf{F}_p^V$; moreover, this correspondence neatly places the translation planes in their proper place according to the size of their kernels (see Proposition 2 below). The codes $B(\mathbf{F}_q|F)$ are the *standard choices*.

EXAMPLES:

(1) For $q = 9$ the only interesting standard choice is $B(\mathbf{F}_9|\mathbf{F}_3)$; it is an [81,50] ternary code with 2×1170 weight-9 vectors. $B(\mathbf{F}_9|\mathbf{F}_3)$ is code-isomorphic to $Hull(\Omega^T)^\perp$ where Ω is the non-desarguesian translation plane of order nine and T its translation line. $Hull(\Omega^T)$ has many vectors of weight 18 that are not of the form $v^\ell - v^m$ where ℓ and m are parallel lines. One cannot, therefore, recover Ω from its hull.

(2) For $q = 16$ we have $\mathbf{F}_2 \subset \mathbf{F}_4 \subset \mathbf{F}_{16}$. Thus,
$$C(AG_2(\mathbf{F}_{16})) = B(\mathbf{F}_{16}|\mathbf{F}_{16}) \subset B(\mathbf{F}_{16}|F_4) \subset B(\mathbf{F}_{16}|\mathbf{F}_2)$$
and the dimensions of these three codes are 81, 129 and 163 respectively. There are precisely three translation planes linear over $B(\mathbf{F}_{16}|\mathbf{F}_4)$: see [**14**]. The codes of the two non-desarguesian planes each have dimension 97. It is not known whether or not they are linearly equivalent (see Definition 2 below).

PROPOSITION 2. *Let q be a power of the prime p. Then an affine plane of order q is linear over $B(\mathbf{F}_q|\mathbf{F}_p)$ if and only if it is a translation plane. Moreover, if it is contained in $B(\mathbf{F}_q|F)$ then its kernel contains F and its hull is contained in $E(\mathbf{F}_q|F)$.*

Proposition 2 will yield, immediately, upper bounds on the dimensions of the codes of affine translations planes. Roughly speaking, as the kernel of the translation plane gets larger, the dimension gets smaller and, moreover, one can prove a weak version of the Hamada-Sachar Conjecture. First, two further definitions:

DEFINITION 2. *Let $\mathcal{D}$ and $\mathcal{E}$ be two designs with the same parameters; we take their point sets to be the same set $\mathcal{P}$. The designs are said to be linearly equivalent (or, more properly, "linearly equivalent at p") if $Hull_p(\mathcal{D})$ is code-isomorphic to $Hull_p(\mathcal{E})$. That is, if there is an element σ of $Sym(\mathcal{P})$ acting naturally on $\mathbf{F}_p^{\mathcal{P}}$ with $\sigma(Hull_p(\mathcal{D})) = Hull_p(\mathcal{E})$.*

DEFINITION 3. *A projective plane Π of order n is said to be tame (or, more properly, "tame at p") if $Hull_p(\Pi)$ has minimum weight $2n$ and the minimal-weight vectors are precisely the scalar multiples of the vectors of the form $v^L - v^M$, where L and M are lines of the plane.*

REMARK:

All desarguesian planes are tame, but not all planes are; for example Ω, the non-desarguesian translation plane of order nine, is not (as Example 1 above indicates).

THEOREM 3. *Suppose π is a translation plane contained in $B(\mathbf{F}_q|F)$ where F is a subfield of $\mathbf{F}_q$. Then*

$$dim(C(\pi)) \leq q + dim(E(\mathbf{F}_q|F)).$$

Moreover, any two translation planes contained in $B(\mathbf{F}_q|F)$ meeting this bound are linearly equivalent.

A proof of Theorem 3 can be found in [5]. Although there is not, in general, an easy to use formula for the dimension of the code $E(\mathbf{F}_q|F)$ there are such formulae in special cases. The reader should consult [3] and [13].

Unfortunately, I do not have an example of two different affine translation planes meeting the bound nor of two different linearly equivalent planes. There are, however, different designs that are linearly equivalent in a non-trivial way: the Ree and Hermitian unitals on 28 points are linearly equivalent, but different. Denoting the Hermitian unital by $\mathcal{H}$ and the Ree unital by $\mathcal{R}$, then $C_p(\mathcal{H})$ and $C_p(\mathcal{R})$ will both be taken in $\mathbf{F}_p^{\mathcal{P}}$, $\mathcal{P}$ being the point set of each. These codes are $\mathbf{F}_p^{\mathcal{P}}$ whenever $p \neq 2$. But $C_2(\mathcal{H})$ is a $[28,21]$ binary code and $C_2(\mathcal{R})$ is a $[28,19]$ binary code. Each code has minimum weight 4, but $C_2(\mathcal{H})$ has 315 weight-4 vectors whereas those of $C_2(\mathcal{H})$ are simply the 63 blocks of the Ree unital. These *codes* look rather different, but the two *hulls*, $Hull_2(\mathcal{H}) = C_2(\mathcal{H}) \cap C_2(\mathcal{H})^{\perp}$ and $Hull_2(\mathcal{R}) = C_2(\mathcal{R}) \cap C_2(\mathcal{R})^{\perp}$, are isomorphic $[28,7]$ binary codes with weight distribution

$$x^0 + 63(x^{12} + x^{16}) + x^{28}$$

and thus the designs are linearly equivalent at 2. In fact, by a proper choice of the blocks in $\mathcal{P}$ one can arrange matters so that one has the following Hasse diagram:

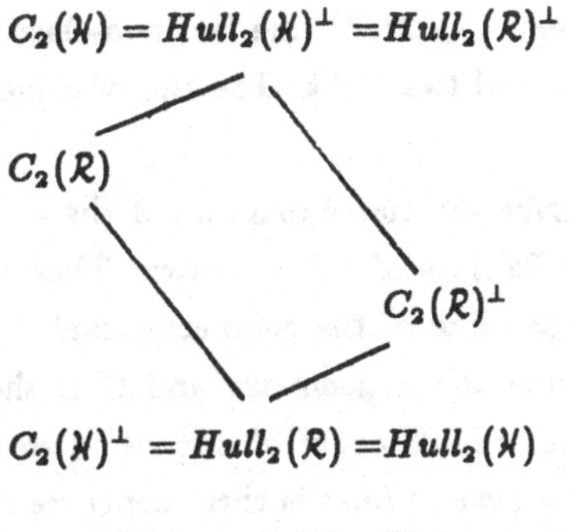

Observe that, in this case, $C_2(\mathcal{H})^{\perp} \subset C_2(\mathcal{H})$. The automorphism group of $C_2(\mathcal{H})$ is much larger than that of $\mathcal{H}$: it is the symplectic group $Sp_6(\mathbf{F}_2)$. Put another way, both $\mathcal{H}$ and $\mathcal{R}$ can be extracted from a larger design with parameters $N = 28, d = 4, \lambda = 5$ and given by the supports of the minimal-weight vectors of $C_2(\mathcal{H})$. This design's automorphism group is $Sp_6(\mathbf{F}_2)$ in its doubly-transitive action on 28 points. The phenomenon just described, viz extracting designs with $\lambda = 1$ from a design with a larger λ, is very much akin to the production of a non-desarguesian plane from a given plane via the process of derivation, the subject of the next section. Frequently one starts with the best design, the desarguesian plane, and extracts many others. Of course, it takes considerable ingenuity to extract these designs. In the case at hand it can be done easily via group theory.

An even more striking example of this phenomenon can be seen using the three $(16, 6, 2)$ designs, where one has an especially good design to play the rôle of the desarguesian plane. This design has a doubly-transitive automorphism group, its code and its hull are identical, and the minimal-weight vectors are the blocks of the design, but the *orthogonal* to the hull has many more vectors of weight six (in fact, 256 forming a two-design with $\lambda = 32$) and amongst these vectors the other two designs may be found [6]. Put another way this design is linear over all designs with parameters (16,6,2).

This last example treated the biplanes of order four. J. D. Key and I have recently given a similar treatment to the five triplanes of order four. Computer calculations using CAYLEY on Clemson University's VAX led to the following rather concise description. The parameters in question are (15,7,3) and again there is a classical design to play the rôle of the desarguesian plane: the design of points and planes in the three-dimensional projective geometry over $\mathbf{F}_2$. The code of this design is of dimension five and B, the orthogonal to the hull, is the [15,11] binary Hamming code. The weight-7 vectors in B, 435 in number, are a design with $\lambda = 87$ and all five triplanes are subdesigns of this design; i.e., the classical design is linear over all designs with parameters (15,7,3). Of the four non-classical designs one code has dimension six, one seven, and two eight. The two of dimension eight are duals of one another.

It is very easy to describe the Hasse diagram of the codes involved in terms of the geometry—using the 35 lines of the geometry. There are 35 distinct codes of dimension six, one for each line of the geometry; each such code is obtained as $C \oplus \mathbf{F}_2 v^{\ell}$ where ℓ is a line of the geometry and C is the code of the design of points and planes. There are 105 codes of dimension seven, one for each planar 3-claw of lines (a *planar 3-claw of lines* is three concurrent lines lying in a plane); each such code is obtained as $C \oplus \mathbf{F}_2 v^{\ell} \oplus \mathbf{F}_2 v^{m}$, where ℓ and m are two lines of the claw. Finally, there are 15 distinct codes of dimension eight for each of the two different designs. One type is obtained as $C + H$, where H is the code of a Fano plane of the geometry, and the other as $C + K$, where K is generated by the supports of a complete 7-claw of lines. The containment relations are the obvious ones. There are, of course, more designs amongst the weight-7 vectors of B; what we are saying is that each of the codes we have described is generated by one of the appropriate designs and every design whose code contains C generates exactly one of these codes.

The ideas in the above example have analogues in higher dimensions and it is possible that a coherent account—at least up to linear equivalence—of all designs with the parameters of the design of points and hyperplanes of projective m-space over $\mathbf{F}_2$ can be had using them.

In order to prove the weak version of the Hamada-Sachar conjecture one must have a criterion that tells one when linearly equivalent affine planes are, in fact, isomorphic. Because of the fact that the objects of interest are projective planes, the criterion is given in those terms.

THEOREM 4. *Suppose* Π *and* Σ *are projective planes of order* n *and that* Π *is tame at* p, *where* p *is odd. If for some line* L *of* Π *and some line* M *of* Σ, Π^L *is linearly equivalent at* p *to* Σ^M, *then* Π *and* Σ *are isomorphic.*

It follows from the work of Delsarte et al (see [**3**]) that desarguesian projective planes are tame. Thus Theorem 4 allows one to identify a plane that is desarguesian: a projective plane is desarguesian provided that is has an affine part linearly equivalent to a desarguesian affine plane. Conjecturally, a desarguesian plane can be identified by the dimension of its code. We can prove a weak result in that direction; Theorem 4 is crucial to the proof of the following theorem (the complete proof can be found in [**3**]).

THEOREM 5. *Let* π *be an affine plane of odd order* $q = p^s$ *contained in* $B(\mathbf{F}_q|\mathbf{F}_p)$. *Then, if* $C(\pi)$ *contains a subcode isomorphic to* $Hull(AG_2(\mathbf{F}_q))$, *we have that*

$$\dim C(\pi) \geq \binom{p+1}{2}^s$$

with equality only if π *is isomorphic to* $AG_2(\mathbf{F}_q)$.

This Theorem is weak not so much because of the condition of containment in $B(\mathbf{F}_q|\mathbf{F}_p)$ (since at least all translation planes are captured) but because of the assumption that $C(\pi)$ contains a subcode isomorphic to the hull of the desarguesian plane. We have not been able to see how to eliminate this assumption. The restriction to *odd order* may be necessary. It is not at all clear that the Hamada-Sachar Conjecture should be true for planes of order 2^m when $m > 3$ (see Remark (2) to Theorem 1).

DERIVATIONS

The notion of a *derivation* finds a natural setting here. One needs a rather technical result from [3] to properly discuss the notion. At hand is an affine plane π of order n; Π will be its projective completion with L_∞ the line at infinity. Fix a prime p dividing n and set $H = Hull_p(\pi)$ and $B = H^\perp$. B has a minimum weight n and all minimal-weight vectors are constant; further, amongst the minimal-weight vectors are the scalar multiples of the $n^2 + n$ lines of π. In order to investigate the nature of those minimal-weight vectors of B which are not scalar multiples of lines of π, one relates these vectors to their preimages, under the natural projection, in $Hull_p(\Pi)^\perp$. Set $\bar{B} = Hull_p(\Pi)^\perp$ and let $T : \bar{B} \longrightarrow B$ be the natural projection. Here is the technical result:

FUNDAMENTAL LEMMA. ***Let b be a minimal-weight vector in B. Then, for p odd, there is a unique constant vector $\bar{b}$ in $\bar{B}$ with $T(\bar{b}) = b$. Moreover $wgt\ \bar{b} = n+r$ with $r \equiv 1(\text{mod}\ p)$, $r < n$; r is the number of parallel classes in π for which the support of b is not a transversal. When $p = 2$ there are two such vectors and uniqueness in achieved by requiring that $\bar{b}$ be not in $C_2(\Pi)^\perp$.***

Now let Π be an arbitrary projective plane of order n and p a prime dividing n. Let L be any line of Π. For any subset D of L one can consider the affine plane Π^L and the vectors $b \in B$ whose unique preimages in $\bar{B}$ (as in the Lemma) have a support whose intersection with L is precisely D. Thus, L plays the rôle of L_∞ in the Fundamental Lemma. Here now is the definition of what might better be called a *primitive derivation set*.

DEFINITION 4. ***Let D be a subset of points of a line L of a finite projective plane Π of order n. D is said to be a derivation set for Π if there is a prime p dividing n and a collection $\mathcal{D}$ of vectors of $\bar{B}$ satisfying the following conditions:***

(1) $|\mathcal{D}| = n|D|$,

(2) *each $\bar{b} \in \mathcal{D}$ is of the form v^X where $D \subset X$ and $wgt(\bar{b}) = n + |D|$,*

(3) *for distinct $\bar{b}$ and $\bar{c}$ in $\mathcal{D}$, $wgt(\bar{b} - \bar{c}) \geq 2n - 2$.*

The point of the Definition is that using Π and $\mathcal{D}$ it is quite trivial to construct an affine plane as follows: the points are the points of $\Pi^L = \pi$ and the lines are those of π coming from lines of Π that don't meet D together with those subsets X of π's points coming from the vectors v^X that are images of the vectors $\bar{b} \in \mathcal{D}$. For a full account the reader should consult [5].

Theorem 2 gives immediately the following result:

PROPOSITION 3. *For a plane of prime order there do not exist non-trivial derivation sets. For a plane whose order is the square of a prime the only possible non-trivial derivation sets are Baer segments.*

A *Baer segment* of a line of a projective plane is the intersection of that line of the projective plane with a Baer subplane where the intersection contains at least two points. The classical derivations (see [**16**]) use Baer segments but other choices are possible (see [**5**]).

CONCLUSION

The traditional approaches to the theory of designs are either through group theory or through set theory proper. It appears that most designs are rigid (i.e., have no automorphisms at all) and hence the group-theoretical approach, at least as currently practiced, seems best suited to classification. The recursive constructions of set theory, on the other hand, have been the most powerful as far as existence is concerned (as is witnessed by the stunning result of Teirlinck), but ill-suited to classification. The ideas presented here occupy a middle ground and it is too early to say just what their future will be. It must be said, however, that there are designs, even interesting ones with doubly-transitive groups, that the theory sketched above will not treat. An example is the design of ovals in an odd-order desarguesian projective plane; in this instance no matter which prime one chooses neither the code nor the hull give information (see [**4**]).

The point is that to give information the hull must be non-trivial and must preferably be capable of producing related designs. The three most interesting examples known at present are the $(16, 6, 2)$ design with the doubly-transitive automorphism group, the Ree unital on 28 points, and the design of hyperplanes in the three-dimensional projective geometry over $\mathbf{F}_2$. I have no doubt that others will be found and there is a great deal of room for investigation, via the above ideas, of existing designs.

The theory will also relate designs with the same number of points, N, but different block sizes, d. For example, the affine plane of order four and the three biplanes of order four are closely related via the $[16, 5]$ extended Reed-Muller code and, in turn, these are related, via the [15,4] Reed-Muller code, to the five triplanes of order four.

APPENDIX

ADMISSIBLE PARAMETERS FOR DESIGNS

The appendix is independent of the coding theory aspects of the paper and therefore the usual notation of design theory will be used. Only so-called *two-*

designs are treated although t-designs can be treated in much the same way. So a *design*, $(\mathcal{P}, \mathcal{B})$, will be a point set $\mathcal{P}$ together with a block set $\mathcal{B}$ with a bipartite graph structure imposed, $\mathcal{P}$ and $\mathcal{B}$ being the parts. Each point will have valency r and each block valency k; the path condition, no longer symmetric, will remain the same: the number of paths of length two from any one point to any other will be independent of the two points chosen and this number will be denoted by λ. The cardinality of $\mathcal{P}$ will be denoted by v and that of $\mathcal{B}$ by b —as is customary. By Fisher's inequality $v \leq b$. Observe that we have not eschewed repeated blocks. The two congruence constraints, namely

$$\lambda v(v-1) = bk(k-1) \text{ and } rv = bk$$

are simply counts of the number of paths of length two between two points and the number of edges, made, as usual, in the two possible ways. It follows that

$$\lambda(v-1) = r(k-1). \qquad *$$

Let $n = r - \lambda$ denote the *order* of the design. Observe that * implies that $n(k-1)$, clearly congruent to $r(k-1)$ modulo λ, is congruent to 0 modulo λ and setting $n(k-1) = \lambda\mu$ one has that $v = k + \mu$. Moreover, since b must be an integer, a necessary condition on n, λ, and μ is that

$$\boxed{\mu(n+\lambda) \equiv 0 \text{ (modulo } k)}$$

with $k \leq n + \lambda$ because of Fisher's inequality. This single congruence constraint replaces two because of our introduction of the parameter μ. One now has two choices: fix k and λ and allow n to vary taking account of the constraint or fix n and λ and examine the finitely many k's. Although I prefer the second course there is a place for the first. An example of the first procedure is afforded by Steiner triple systems: the parameters are $(2n+3, 3, 1)$ with $n \equiv 0$ or 2 (modulo 3).

For Steiner systems (i.e. for $\lambda = 1$) the necessary condition becomes

$$n(n+1) \equiv 0 \text{ (modulo } k), \ k \leq n+1$$

since, in this case, $\mu = n(k-1) \equiv -n$ (modulo k). The parameters are

$$(n(k-1)+k, k, 1).$$

Fixing $\lambda = 2$, the necessary condition becomes:

$$\frac{n}{2}(n+2) \equiv 0 \text{ modulo } k, \ k \leq n+2, \text{ for } n \text{ even}$$

and

$$n(n+2) \equiv 0 \text{ modulo } k, \ k \leq n+2, \text{ for } n \text{ odd.}$$

A table is given below, where the row is the order and the column the λ. Boldface indicates the presence of a classical or neo-classical design and italics non-existence because of the Bruck-Ryser-Chowla Theorem. For Steiner systems (the first column) the parameter sets at each order begin with the design of all two-subsets of an $(n+2)$-set and the last two parameter sets are those of the affine plane and the projective plane of the given order. The first time the Bruck-Ryser theorem is invoked is at order six and through order seven all Steiner Systems that could exist do exist. The first question mark would come at order eight: it is not known whether or not there is a design with parameters $(46, 6, 1)$. (The Steiner systems with $k = 6$ afford another instance where it would be appropriate to fix k and vary n. Here the parameters are $(5n + 6, 6, 1)$ with $n \equiv 0$ or 2 (modulo 3) and the existence question is almost settled.) In the second column (i.e. for $\lambda = 2$) the Bruck-Ryser-Chowla Theorem is first invoked at order five (there are no designs of order five with $\lambda = 2$). If repeated blocks are allowed existence becomes easier. In the case at hand, for example, when n is even the first design exists when repeated blocks are allowed, but won't if they are not.

Once again, one need not recursively generate a table entry; for any fixed λ and n one can simply list and examine the possible parameter sets. Again, order twelve is interesting; we leave the examination to the reader.

Observe that there are no parameter sets for $\lambda = 3$ and $n = 2$, for $\lambda = 4$ and $n = 3$, and for $\lambda = 5$ and $n = 2$. A defect of the above tabularization is that the complement of a design, although it appears in the same row, is in a different column. Taking complements amounts to interchanging the rôles of k and μ; thus, the λ of the complement is $\frac{n(\mu-1)}{k} = \frac{\mu(n+\lambda)}{k} - n$. For example, for order four, the complement of (10,4,2) is (10,6,5). Perhaps another table that holds $k + \mu$ fixed is in order.

Table of admissible parameters

	1	2	3	4	5
1	(3,2,1)	(4,3,2)	(5,4,3)	(6,5,4)	(7,6,5)
2	**(4,2,1)** **(7,3,1)**	(3,2,2) **(7,4,2)**		(4,3,4)	
3	(5,2,1) **(9,3,1)** **(13,4,1)**	(6,3,2) **(11,5,2)**	(3,2,3) (5,3,3) **(11,6,3)**		(9,6,5)
4	(6,2,1) **(16,4,1)** **(21,5,1)**	(4,2,2) (7,3,2) (10,4,2) **(16,6,2)**	**(8,4,3)** **(15,7,3)**	(3,2,4) (7,4,4) **(15,8,4)**	(10,6,5)
5	(7,2,1) (13,3,1) **(25,5,1)** **(31,6,1)**	*(15,5,2)* *(22,7,2)*	(9,4,3)	(10,5,4) (19,9,4)	(3,2,5) (9,5,5) (19,10,5)
6	(8,2,1) (15,3,1) *(36,6,1)* *(43,7,1)*	(5,2,2) (9,3,2) (13,4,2) *(21,6,2)* *(29,8,2)*	(4,2,3) (7,3,3) (16,6,3) (25,9,3)	(6,3,4) (11,5,4)	(12,6,5) (23,11,5)

REFERENCES

1. E. F. Assmus, Jr., *The coding theory of finite geometries and designs*, Springer Lecture Notes in Computer Science (to appear).
2. E. F. Assmus, Jr. and J. D. Key, *Arcs and ovals in the hermitian and Ree unitals*, European J. Combin. (to appear).
3. ———, *Affine and projective planes*, Discrete Mathematics (to appear).
4. ———, *Baer subplanes, ovals and unitals*, Coding Theory, IMA Volumes in Mathematics and Its Applications, ed. D. Ray-Chaudhuri, **20** (1989), Springer-Verlag.
5. ———, *Translation planes and derivation sets*, IMA preprint series (to appear).
6. E. F. Assmus, Jr. and C. J. Salwach, *The (16,6,2) designs*, Internat. J. Math.& Math. Sci., **2** (1979), 261–281.

7. H. Beker and W. Haemers, *2-designs having an intersection number $k-n$*, J. Combin. Theory, Ser. A **28** (1980), 64–81.

8. P. Delsarte, *A geometric approach to a class of cyclic codes*, J. Combin. Theory, **6** (1969), 340–358.

9. P. Delsarte, J. M. Goethals and F. J. MacWilliams, *On generalized Reed-Muller codes and their relatives*, Information & Control 16, **5** (1970), 403–442.

10. R. H. F. Denniston, *Subplanes of the Hughes plane of order 9*, Proc. Camb. Phil. Soc. **64** (1968), 589-598.

11. J. F. Dillon and J. R. Schatz, *Block designs with the symmetric difference property*, Proceedings of the NSA Mathematical Sciences Meetings, 6-7 January 1987 and 7-8 October 1987. Edited by Robert L. Ward.

12. W. H. Kantor, *Exponential numbers of two-weight codes, difference sets and symmetric designs*, Discrete Math. **46** (1983), 95–98.

13. J. D. Key and K. Mackenzie, *An upper bound for the p-rank of a translation plane (submitted).*

14. E. Kleinfeld, *Techniques for enumerating Veblen–Wedderburn systems*, J. Assoc. Computing Machinery **7** (1960), 330-337.

15. C. W. H. Lam, T. Thiel, S. Swiercz and J. McKay, *The nonexistence of ovals in a projective plane of order 10*, Discrete Math. 45 (1983), 319–321.

16. T. G. Ostrom, *Finite Translation Planes*, Lecture Notes in Mathematics, **158**, Springer (1970).

17. C. J. Salwach, *Planes, biplanes and their codes*, Amer. Math. Monthly **88** (1981), 106–125.

18. Tran van Trung, *A symmetric design with parameters (61,16,4)*, J. Combin. Theory, Ser. A **37** (1984), 374.

Department of Mathematics
Lehigh University
Bethlehem, Pennsylvania 18015 U.S.A.

Designs: mappings between structured sets

R. A. Bailey

1 Structured sets

Every statistical design consists of two sets and a function between them. One set, $\mathcal{T}$, consists of the treatments: they are under the experimenter's control, and the purpose of the experiment is to find out about them. The elements of the second set, Ω, are called *plots* for historical reasons. A plot is the smallest experimental unit to which an individual treatment is applied: it may or may not be a plot of land. In general the experimenter has less direct control over the attributes of the plots, and is not interested in finding out about the plots per se. Unless otherwise stated, Ω and $\mathcal{T}$ are always finite. Finally there is the design map ϕ from Ω to $\mathcal{T}$ which allocates treatments to plots; if treatment t is allocated to plot ω then $\omega\phi = t$.

Typically one or both of the sets $\mathcal{T}$ and Ω is *structured*. I cannot give a formal definition of this concept, but examples of structures on a set include: a set of partitions; a transitive permutation group; a set of subsets; a graph; an association scheme. Many, but not all, of these structures can be specified as a set of (binary) relations with certain properties. A set will be said to be *unstructured* if the relations in its structure are just the equality relation E and the uncaring relation U (the direct square of the whole set) or $U \setminus E$.

Classes of design are defined by the way that ϕ links the structure on Ω to that on $\mathcal{T}$. If R and S are relations on Ω and $\mathcal{T}$ respectively, ϕ will be said to *satisfy* $[R\,;S]$ if there is an integer $n(R,S)$ such that

$$(t,u) \in S \quad \Rightarrow \quad |\{(\alpha,\beta) \in R : \alpha\phi = t,\ \beta\phi = u\}| = n(R,S).$$

In particular, if ϕ satisifies $[E;E]$ then ϕ is *equireplicate* with replication $n(E,E)$, which is equal to $|t\phi^{-1}|$ for all t in $\mathcal{T}$.

Here are some familiar examples expressed in these terms.

EXAMPLE 1 (BLOCK DESIGNS) Suppose that the structure on Ω is a partition into blocks. Let B be the corresponding equivalence relation on Ω. Every design $((\Omega,B),\mathcal{T},\phi)$ is a block design, which is *proper* if the blocks have equal size. Because blocks are subsets of Ω, not of $\mathcal{T}$, there is no problem about allowing two different blocks to be allocated the same set of treatments, or allowing a treatment to occur more than once in a block. Now $(\Omega,\mathcal{T},\phi)$ is a *complete-block* design if the restriction of ϕ to each block is a bijection; that is, if ϕ satisfies $[E\,;E]$ and $[B\,;U]$ with $n(B,U) = n(E,E)$.

For an *incomplete-block* design the restriction of ϕ to each block is one-one; that is, ϕ satisfies $[B\backslash E;E]$ with $n(B\backslash E,E) = 0$: furthermore, the image of each block is a proper subset of $\mathcal{T}$ (hence the word *incomplete*). A proper block design is *balanced* if it is equireplicate and satisfies $[B;E]$ and $[B;U\setminus E]$, whether or not it has incomplete blocks. If the block size is k and ϕ induces a bijection between the blocks and the k-subsets of $\mathcal{T}$ then $(\Omega,\mathcal{T},\phi)$ is a balanced incomplete-block design according to the statistical definition. Many pure mathematical texts explicitly exclude this case. □

EXAMPLE 2 (LATIN SQUARES) Now let Ω be a square array, with equivalence relations R (for rows) and C (for columns). Then $(\Omega,\mathcal{T},\phi)$ is a Latin-square design if the restriction of ϕ to each row and each column is a bijection. In our present language, ϕ satisfies $[E;E]$, $[R;U]$ and $[C;U]$ with $n(E,E)=n(R,U)=n(C,U)$. This is a very cumbersome way of describing a Latin square, but it does emphasize the difference between the combinatorial object called a Latin square (which I usually describe as a set with three partitions), which can be used to construct many different designs, and the Latin-square *design*, in which Ω and $\mathcal{T}$ haye the roles prescribed above.

Suppose that Ω has further relations $N_1,\ldots,N_4$, where $(\alpha,\beta)\in N_1,\ldots,N_4$ if β is the immediate north, south, east, west neighbour of α respectively. A Latin-square design satisfying $[N_i;U\setminus E]$ for $i=1,\ldots,4$ is said to be *complete*. There are weaker variants of this idea: for example, a square satisfying $[N_i;U\setminus E]$ for $i=3$ and 4 is *row-complete*, while a square satisfying $[N_1\cup N_2;U\setminus E]$ and $[N_3\cup N_4;U\setminus E]$ is *quasi-complete* [45]. □

To the statistician, the design $(\Omega,\mathcal{T},\phi)$ is only half the story. The rest is provided by the real vector spaces $\mathbf{R}^\Omega$ and $\mathbf{R}^\mathcal{T}$, which consist of all functions from Ω (respectively $\mathcal{T}$) to $\mathbf{R}$. Although the experimenter wants to find out something about elements of $\mathbf{R}^\mathcal{T}$ (specifically, to estimate a fixed unknown element v_0 of $\mathbf{R}^\mathcal{T}$), the available data (after one measurement has been made on each plot) is a vector in $\mathbf{R}^\Omega$. The link between these two spaces is provided by the linear transformation $\phi_*\colon \mathbf{R}^\mathcal{T}\to\mathbf{R}^\Omega$, induced by ϕ by the rule

$$(\omega)(v\phi_*)=(\omega\phi)v \qquad \text{for } v \text{ in } \mathbf{R}^\mathcal{T} \text{ and } \omega \text{ in } \Omega.$$

Thus ϕ_* is a kind of dual of ϕ.

Each of $\mathbf{R}^\mathcal{T}$ and $\mathbf{R}^\Omega$ has a natural inner product: the characteristic functions of the singletons form an orthonormal basis. The combinatorial structures on $\mathcal{T}$ and Ω often determine orthogonal direct-sum decompositions of $\mathbf{R}^\mathcal{T}$ and $\mathbf{R}^\Omega$. The decomposition of $\mathbf{R}^\mathcal{T}$ is used in interpreting the results of the experiment (the projection of v_0 onto a space W in the decomposition is known as the *effect* of W), while that of $\mathbf{R}^\Omega$ is used in analysing the experiment—that is, obtaining results from the raw data. It turns out that many of the *combinatorial* conditions on ϕ, linking the structures on $\mathcal{T}$ and Ω, are specified purely to ensure that ϕ_* have desirable *algebraic* or *geometric* properties in the way it links the orthogonal decompositions of the two spaces. Much of the fascination of the pure theory of statistical design comes from this interplay between the combinatorics and the algebra: the connection between the pure theory and the constraints and needs of real experiments is another interesting story, but one into which I shall not go here.

My lack of a precise definition of *structure* means that I cannot pursue the connection between the combinatorics and the algebra at a general level. Instead, in the following sections I shall use this framework to describe several important areas of statistical design. Along the way I shall pose several problems which I hope combinatorialists may have the tools to solve.

2 Fractional factorials

If the structure on $\mathcal{T}$ is provided by a transitive Abelian group then we may identify $\mathcal{T}$ with that group. Its dual group $\mathcal{T}^*$ consists of the irreducible characters of $\mathcal{T}$; that is, all homomorphisms from $\mathcal{T}$ to the non-zero complex numbers. The elements of $\mathcal{T}^*$ are an orthogonal basis for $\mathbf{C}^{\mathcal{T}}$, and hence give an orthogonal direct-sum decomposition of $\mathbf{C}^{\mathcal{T}}$ into 1-dimensional subspaces. Pairing each character with its complex conjugate gives an orthogonal direct-sum decomposition of $\mathbf{R}^{\mathcal{T}}$ into 1- and 2-dimensional spaces: see [63].

In a classical *fractional factorial* design, Ω is also an Abelian group and ϕ is a group monomorphism. There is thus a genuine dual map $\phi^*: \mathcal{T}^* \to \Omega^*$, which is an epimorphism. Extension of ϕ^* by linearity followed by restriction of scalars gives ϕ_*. However, it is easier to work directly with ϕ^*.

If ϕ is not an isomorphism, then there are some treatments which do not occur on any plot. Correspondingly, ϕ^* is many-one and has non-trivial kernel. The elements of $\ker \phi^*$ are called *defining contrasts* of the design; elements of $\mathcal{T}^*$ in the same coset of $\ker \phi^*$ are said to be *aliased* with each other. If χ is in $\ker \phi^*$ then $\chi\phi^*$ is a constant vector and the design gives no information about the effect of χ; if $\chi\phi^* = \eta\phi^*$ then information about the effect of χ cannot be disentangled from that of η.

For a *factorial* design, $\mathcal{T}$ has distinguished subgroups $\mathcal{T}_1, \ldots, \mathcal{T}_N$ such that $\mathcal{T}$ is the internal direct product

$$\mathcal{T} = \mathcal{T}_1 \otimes \cdots \otimes \mathcal{T}_N. \tag{1}$$

I call an Abelian group with such a distinguished decomposition a *named* group. Then $\mathcal{T}^* \cong \mathcal{T}_1^* \otimes \cdots \otimes \mathcal{T}_N^*$, and each χ in $\mathcal{T}^*$ has a *weight* defined to be the number of non-trivial coordinates with respect to this decomposition of $\mathcal{T}^*$. Equivalently, the weight of χ is equal to

$$N - |\{i : 1 \leq i \leq N,\ t\chi = 1\ \forall\, t \in \mathcal{T}_i\}|. \tag{2}$$

The needs of a particular experiment partition $\mathcal{T}^* \setminus \{1\}$ into three sets: *required* effects (those to be estimated), *negligible* effects (those assumed to be zero), and the remainder: see [47]. If ϕ is to be useful then, for every required effect χ, every other character aliased with χ must be negligible. The required effects usually have lower weight than the others. Sometimes the required effects are precisely those characters whose weight is less than some specified value w_0 and the negligible effects are those with weight at least w_1, where $w_1 \geq w_0$. Thus elements of $\ker \phi^*$ should have weight at least $w_0 + w_1 - 1$. Making precise an idea which has been used at least since [24] and [77], we follow [28] and [10] and define:

DEFINITION A subgroup H of T^* is *w-heavy* if every non-trivial element of H has weight at least w. The design ϕ has ***resolution*** w if $\ker \phi^*$ is w-heavy.

EXAMPLE 3 For $i = 1, 2, 3$ let $T_i = \langle a_i \rangle$ with $a_i^5 = 1$, and let T_i^* be generated by A_i, where $a_i A_i$ is a given primitive fifth root of unity in $\mathbf{C}$. Let $\chi = A_1 A_2 A_3$. Then $\langle \chi \rangle$ is 3-heavy. For a resolution 3 design we can choose Ω and ϕ so that $\ker \phi^* = \langle \chi \rangle$. Then

$$\operatorname{Im} \phi = \{a_1^{n_1} a_2^{n_2} a_3^{n_3} : n_1 + n_2 + n_3 = 0 \bmod 5\},$$

which is another manifestation of a Latin square. □

Since ϕ is a monomorphism, the structure of Ω is faithfully reflected in $\operatorname{Im} \phi$, which is just the annihilator of $\ker \phi^*$. Thus the search for a good design becomes the search for a suitable subgroup of T^*. The larger is $\ker \phi^*$ the smaller is Ω and so the cheaper is the experiment.

Problem 1 Given Abelian groups $T_1, \ldots, T_N$ and a natural number w, find the maximal w-heavy subgroups of $T_1 \otimes \cdots \otimes T_N$.

It is clear ([31, 10, 87]) that the solution to this problem is obtained by taking the direct product of the solutions for the Sylow subgroups of T. Hence we may assume that each T_i is a p-group for the same prime p. If each T_i is cyclic of order p then the problem is identical to one in coding theory [25]: find the maximal linear codes over $\mathsf{GF}(p)$ with word-length N and minimum distance w. Thus any results found in coding theory (such as [85]) are of interest to statisticians. But our problem is more general than that of coding theorists, because the T_i are not required to be isomorphic, nor even of the same size, nor even elementary Abelian. In fact, two interesting sub-problems of Problem 1 show this generality more clearly.

Problem 2 When does replacement of T_i by another Abelian group of the same order change the maximal size of w-heavy subgroups of T?

Problem 3 Given an Abelian p-group H and a natural number w, which named p-groups have w-heavy subgroups isomorphic to H?

Problem 2 is not idle. If $\mathbf{C}_n$ and $\mathbf{V}_n$ denote the cyclic and elementary Abelian groups of order n respectively, then the named group $\mathbf{C}_2 \times \mathbf{C}_2 \times \mathbf{C}_4 \times \mathbf{C}_4$ has no 3-heavy subgroup of order 4 but the named group $\mathbf{C}_2 \times \mathbf{C}_2 \times \mathbf{V}_4 \times \mathbf{V}_4$ does have. It has been rather loosely conjectured in [10, 46] that replacing each $\mathcal{T}_i$ by an elementary Abelian group of the same order increases (or at least does not decrease) the maximal size of w-heavy subgroups of $\mathcal{T}$, but nothing has been proved.

Partial answers to Problem 3 are in [10, Theorem 9]. I should welcome a more extensive solution.

3 Factorial structures

In general, a design is said to be *factorial* if $\mathcal{T}$ is a Cartesian product

$$\mathcal{T} = \mathcal{T}_1 \times \cdots \times \mathcal{T}_N \tag{3}$$

with $N \geq 2$ and $|\mathcal{T}_i| > 1$ for $i = 1, \ldots, N$. Put $I = \{1, \ldots, N\}$ and let π_i be the natural projection from $\mathcal{T}$ onto $\mathcal{T}_i$. For each subset J of I, define subspaces V_J and W_J of $\mathbf{R}^{\mathcal{T}}$ by

$$V_J = \left\{ v \in \mathbf{R}^{\mathcal{T}} : tv = uv \text{ if } t\pi_i = u\pi_i \ \forall\, i \in J \right\}; \tag{4}$$

$$W_J = V_J \cap \left(\sum_{K \subset J} V_K \right)^{\perp}. \tag{5}$$

Then $\mathbf{R}^{\mathcal{T}}$ is the orthogonal direct sum of the spaces W_J for $J \subseteq I$. If $J = \{i\}$ then W_J is known as the *main effect of factor i*; otherwise W_J is called the *interaction between the factors in J*. If $K \subset J$ then estimation of the effect of W_K is usually more important than that of W_J, and the former cannot be assumed zero unless the latter is.

Each subset J of I defines an equivalence relation R_J on $\mathcal{T}$ by

$$(t, u) \in R_J \iff t\pi_i = u\pi_i \text{ for all } i \text{ in } J.$$

An alternative definition of V_J is as the set of vectors which are constant on each class of R_J. Thus the dimension of V_J is equal to the number of classes of R_J, which is $\prod_{i \in J} |\mathcal{T}_i|$. A set which is a Cartesian product over I with equivalence relations defined by all subsets of I in this way is a *factorial block structure*.

The structure of Ω may also be a factorial block structure: for example, in a Latin square $\Omega = \Omega_1 \times \Omega_2$ with $|\Omega_1| = |\Omega_2|$. There are various generalizations of factorial block structure which occur more often on Ω than on $\mathcal{T}$. For the first of these, called a *poset block structure* or *distributive block structure*, the set I is assumed to be partially ordered by $\preceq$. Then a subset J of I is *ancestral* (or *admissable* or a *filter* or an *up-set* ...) if

$$i \in J \text{ and } i \preceq j \ \Rightarrow\ j \in J.$$

Then only those R_J for ancestral J are included in the structure. For example, if $\Omega = \Omega_1 \times \Omega_2$ and $|\Omega_1| = b$, $|\Omega_2| = k$ and the poset is the chain $2 \preceq 1$ then we obtain the simple partition into b blocks each of size k.

Since poset block structures arise so often in practice, statisticians have long been fascinated by ***how many*** such structures there might be for a given size of I.

Problem 4 For all N, find the number $\psi(N)$ of isomorphism classes of posets of size N.

Of course, the first few values of $\psi(N)$ may be found in [78, Sequence 588], but no general formula is known. When statisticians first started enumerating ψ, (see [83, 61]) they were unaware that they were counting posets and so simply duplicated Birkhoff's work [23].

Two natural constructions of experimental material—splitting a plot into smaller pieces, and forming all combinations as in an abstract rectangular array—lead to *nesting* and *crossing* of two posets. In the latter no pair from the two posets is comparable; in the former every element of one poset is less than every element of the other. Posets formed from singleton sets by repeated use of either of these constructions are precisely the N-free posets: those where no induced sub-poset is

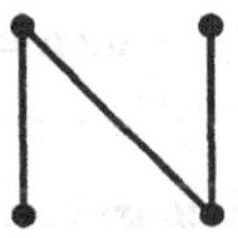

Problem 5 For all N, find the number $\xi(N)$ of isomorphism classes of N-free posets of size N.

Cameron [29] has given a recursive formula for $\xi(N)$, found a bound on the radius of convergence of the power series for ξ, and shown that $\xi(N)/\psi(N) \to 0$ as $N \to \infty$, but I know of no closed formula for $\xi(N)$. In [30] he effectively constructs a graph $\mathcal{G}_N$ whose vertices are the isomorphism classes of N-free posets of size N and whose edges are complementary pairs of N-free posets. Although he finds a closed formula for the number of edges of $\mathcal{G}_N$, the only help this gives to Problem 5 seems to be to place fairly weak bounds on $\xi(N)$.

The set of equivalence relations in a poset block structure satisifies the following conditions:

(A) every equivalence class is *uniform* (that is, has classes of equal size);

(B) the two trivial equivalence relations (E and U) are included;

(C) every pair R, S of equivalence relations is *orthogonal* in the sense that

$$\left(V_R \cap (V_R \cap V_S)^{\perp}\right) \perp \left(V_S \cap (V_R \cap V_S)^{\perp}\right);$$

(D) if R and S are in the structure then so are $R \wedge S$ and $R \vee S$, where (as sets) $R \wedge S = R \cap S$, and $R \vee S$ is the transitive closure of $R \cup S$.

As a further generalization, I define an *orthogonal block structure* to be a set with a family of equivalence relations satisfying (A)–(D). The simplest example of an orthogonal block structure which is not a poset block structure is the Latin square (in another of its guises!): the non-trivial partitions have classes called rows, columns and letters. An orthogonal direct-sum decomposition of $\mathbf{R}^{\mathcal{T}}$ is obtained as before, with the proviso that some of the spaces W may be zero.

Condition (D) shows that the equivalence relations in an orthogonal block structure form a lattice under $\wedge$ and $\vee$. This lattice is known to be Arguesian: that is, it satisfies

$$\begin{aligned}&(R_1 \vee S_1) \wedge (R_2 \vee S_2) \leq R_3 \vee S_3 \quad \Rightarrow \\ &(R_1 \vee R_2) \wedge (S_1 \vee S_2) \leq [(R_1 \vee R_3) \wedge (S_1 \vee S_3)] \vee [(R_2 \vee R_3) \wedge (S_2 \vee S_3)]\end{aligned}$$

But we do not know if this condition is sufficient for a lattice to be represented by the equivalence relations of an orthogonal block structure. Speed and I [80] could not find any Arguesian lattice which could not be so represented.

Problem 6 Find an algebraic characterization of lattices which can be realised as the equivalence relations of an orthogonal block structure.

Groups provide a source of orthogonal block structures. If H is a subgroup of a group G, let R_H be the partition of G into right cosets of H. Then a family of subgroups of G defines an orthogonal block structure (called a *group block structure*) if it is closed under intersection and join, and each pair of subgroups H_1, H_2 in it commutes in the sense that $H_1H_2 = H_2H_1$. In particular, if G is Abelian then every subgroup lattice of G defines a group block structure on G.

It is clear that every poset block structure is a group block structure—simply identify each T_i with any group of the appropriate order. However, not all orthogonal block structures are group block structures: a Latin square not based on a group gives a simple example. However, Tjur [84, page 80] conjectured that every orthogonal block structure is somehow equivalent to a group block structure with an Abelian group.

Problem 7 Is it true that, given an orthogonal block structure, there is an Abelian group with a group block structure such that (i) the lattice of the orthogonal block structure is isomorphic to the lattice of the group block structure and (ii) equivalence relations which correspond under the isomorphism have classes of the same size?

We can now return to strictly factorial designs. When (3) holds, statisticians often identify each T_i with an Abelian group purely for convenience. Then the factorial decomposition $\bigoplus_{J \subseteq I} W_J$ may be further decomposed into the group decomposition $\bigoplus_{H \leq T} W_H$, which in turn may be refined into the character decomposition described in Section 2. Although the first of these decompositions is of primary interest, either of the other two may be used if necessary.

Now, for subgroups H of T, the space V_H is generated (over $\mathbf{C}$) by the characters χ in the annihilator H^0 of H. It is shown in [15] that $W_H = 0$ if H^0 is not cyclic, while otherwise W_H is generated (over $\mathbf{C}$) by the generators of H^0. Further, if $H^0 = \langle \chi \rangle$ then $W_H \leq W_{J(\chi)}$, where

$$J(\chi) = \{i \in I : \chi \notin T_i^0\}$$

(compare this with (2): the weight of χ is equal to $|J(\chi)|$).

In many factorial designs, Ω is a group block structure and ϕ is a group epimorphism (for the particular choices of the T_i): see [63]. Let Γ be a subgroup of Ω whose coset partition contributes to the group block structure. Then [4] shows that, for v in T^*,

$$v\phi_* \in V_\Gamma \iff v \in (\Gamma\phi)^0.$$

Hence, for every subgroup H of T with cyclic annihilator, there is a unique subgroup Γ in the group block structure of Ω such that $W_H\phi_* \leq W_\Gamma$. We say that H is *confounded* with Γ.

For $J \subseteq I$, define the integer $d_{J,\Gamma}$ to be the sum of the dimensions of the spaces W_H for which $W_H \leq W_J$ and $W_H\phi_* \leq W_\Gamma$. Designs (Ω, T, ϕ) and (Ω, T, ϕ') have been defined in [87] to be *degrees-of-freedom equivalent* if, in an obvious notation, $d_{J,\Gamma} = d'_{J,\Gamma}$ for all J and Γ.

Problem 8 For $i = 1, \ldots, N$, suppose given an Abelian group T_i, and let U_i be an elementary Abelian group of order $|T_i|$. Given an Abelian group Ω with a group block structure and given a group epimorphism $\phi : \Omega \to \prod_i T_i$, does there exist a group epimorphism $\phi' : \Omega \to \prod_i U_i$ which is degrees-of-freedom equivalent to ϕ?

The answer to this problem was originally claimed to be "obviously, yes" [70, 46]. However, a proof has been given in [86] *only* for the case that Ω has a single partition into blocks, ϕ is an isomorphism, and the subgroup of treatments in the same block as the identity treatment is cyclic. The general problem now seems rather hard.

4 Strata

The structure on Ω enters the statistical analysis via the assumed covariance matrix for the response y in $\mathbf{R}^\Omega$. This is a symmetric $\Omega \times \Omega$ matrix, $\mathrm{Cov}(y)$, which, loosely speaking, describes how similarly responses on different plots vary. For example, suppose that Ω is a graph with edge-relation R. Identifying subsets of $\Omega \times \Omega$ with their characteristic functions, it is reasonable to assume that

$$\mathrm{Cov}(y) = c_1E + c_2R + c_3U$$

for some scalars c_1, c_2, c_3, probably unknown.

Classical statistical analysis assumes that $\mathrm{Cov}(y)$ is a scalar matrix. However, if W is an eigenspace of $\mathrm{Cov}(y)$ with orthogonal projector P, then $\mathrm{Cov}(yP)$ is a scalar on W and the projected data can be straightforwardly analysed. So long as R is a *regular* graph, the eigenspaces of $c_1E + c_2R + c_3U$ do not depend on the actual values of c_1, c_2, c_3 and so the analysis does not depend on prior knowledge of these coefficients. However, there is a further difficulty. Although the projections of y onto the different eigenspaces are statistically independent, the corresponding eigenvalues are not functionally independent if there are more than three eigenspaces, and so estimation can no longer proceed independently in the separate eigenspaces. To overcome this difficulty, we require that the matrices E, R and U span an algebra over $\mathbf{R}$. This is precisely equivalent to R being a *strongly* regular graph.

EXAMPLE 4 Let Ω consist of the 2-subsets of a 6-set, and let $(\alpha, \beta) \in R$ if and only if $|\alpha \cap \beta| = 1$. Let W_0 be the space of constant vectors in $\mathbf{R}^\Omega$, and let V_1 be the subspace of $\mathbf{R}^\Omega$ spanned by vectors of the form v_i, where i is in the 6-set and

$$\omega v_i = \begin{cases} 1 & \text{if } i \in \omega \\ 0 & \text{otherwise.} \end{cases}$$

Put $W_1 = V_1 \cap W_0^\perp$ and $W_2 = V_1^\perp$. Then W_0, W_1, W_2 are eigenspaces of all matrices of the form $c_1E + c_2R + c_3U$, having dimensions 1, 5 and 9 respectively.

The set Ω may be visualized as the upper half triangle of a square array. All rows and columns are regarded as continuing when reflected in the main diagonal of the square. An edge joins two points if they lie in the same row or column. Now let $\mathcal{T} = \{1, 2, 3, 4, 5\}$ and consider the design ϕ in Figure 1. As always, $W_0 \leq \mathbf{R}^\mathcal{T}\phi_*$; here it may be checked that $\mathbf{R}^\mathcal{T}\phi_* \cap W_0^\perp \leq W_2$. Thus $\mathbf{R}^\mathcal{T} = T_0 \oplus T_2$ where $T_0\phi_* = W_0$ and $T_2\phi_* \leq W_2$. Inference about T_2 is made entirely within W_2. □

Figure 1: Design in Example 4

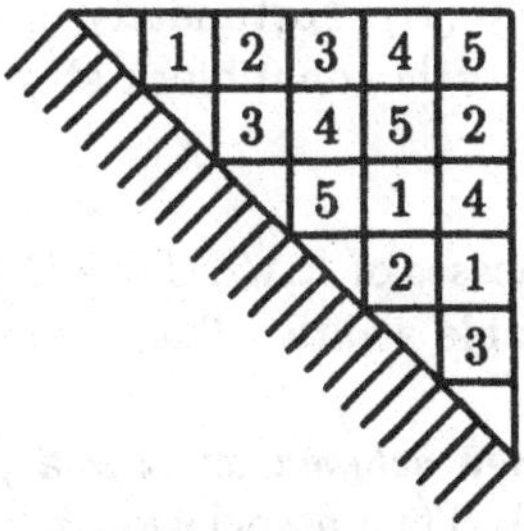

Example 5 If B is a uniform equivalence relation on Ω then $B \setminus E$ is a strongly regular graph. Many mathematical texts exclude this from the definition of strongly regular graph, but it is the one most used in statistical applications. In the notation of Section 3, the common eigenspaces are W_U, W_B and W_E. □

For a structure with more than one non-trivial relation, more than regularity is needed. Let $\mathcal{R}$ be a set of relations on Ω, and suppose that

$$\mathrm{Cov}(y) = \sum_{R \in \mathcal{R}} c_R R \tag{6}$$

for unknown scalars c_R. Then the eigenspaces of Cov(y) are known if and only if

(E) the relations R commute *as matrices*,

(see [88, Sections I.49–50]). Furthermore, these eigenspaces are not too many in number if and only if the matrices R span an algebra over $\mathbf{R}$: that is,

(F) there exist scalars d_{QRS} for Q, R, S in $\mathcal{R}$ such that

$$RS = \sum_{Q \in \mathcal{R}} d_{QRS} Q.$$

Definition Let $\mathcal{R}$ be a set of symmetric relations on Ω. Then the structure $(\Omega, \mathcal{R})$ has *identifiable strata* if the matrices R in $\mathcal{R}$ span a commutative algebra over $\mathbf{R}$: in this case the common eigenspaces of the algebra are called *strata*.

Since covariance matrices are symmetric, there is no need to extend this definition to non-symmetric relations. Moreover, for symmetric relations condition (F) implies condition (E).

Various methods have been proposed for estimating the coefficients c_R, which is a necessary part of analysing the data. For structures with identifiable strata the methods are all equivalent; for structures satisfying (E) but not (F) there is some disagreement among the methods; while there is hardly any agreement for structures which do not satisfy (E). Fortunately, many common structures do have identifiable strata. In an orthogonal block structure

$$RS = k_{R\wedge S} R \vee S$$

where $k_{R\wedge S}$ is the size of the classes of $R\wedge S$. Thus (F) is satisfied and orthogonal block structures have identifiable strata. The strata are the spaces W defined in Section 3: see [80].

By definition, an ***association scheme*** on Ω is a partition $\mathcal{R}$ of Ω into symmetric relations satisfying (F). Thus association schemes also have identifiable strata—the common eigenspaces of the Bose-Mesner algebra [26]. Indeed, orthogonal block structures may be regarded as association schemes, for, if $\mathcal{S}$ is an orthogonal block structure on Ω then there is an association scheme $\mathcal{R}$ on Ω which spans the same algebra: see [80].

Speed [79] has argued that the statistical theory of the analysis of variance requires the structure on Ω to be an association scheme. However, other statisticians do not see the need to require $\mathcal{R}$ to be a *partition* of $\Omega \times \Omega$. Furthermore, there is a different kind of generalization of orthogonal block structure, which is not defined in terms of relations but which has identifiable strata.

A partition of a set Ω may be considered as an equivalence relation or as a set of subsets. The former point of view extends naturally to sets of relations, in particular to association schemes. Here I shall develop the second point of view. Let S be a set of subsets of Ω, each of size k_S, such that every element of Ω is in exactly r_S sets in S. It is tempting to call S a 'block design', but that is incompatible with the other terminology of this paper, so I call it a ***set-cover*** of Ω. The $\Omega \times \Omega$ concurrence matrix C_S of S is defined in the obvious way:

$$(\alpha,\beta)C_S = |\{s \in S : \{\alpha,\beta\} \subseteq s\}|.$$

Now let $\mathcal{S}$ be a set of set-covers of Ω. The statistical reasoning [84, page 51] which suggests that (6) holds for a structure defined by uniform equivalence relations gives

$$\mathrm{Cov}(y) = \sum_{S\in\mathcal{S}} c_S C_S$$

for the structure $(\Omega, \mathcal{S})$. I therefore provisionally define:

DEFINITION A ***stratum geometry*** on Ω is a set $\mathcal{S}$ of set-covers of Ω such that there exist scalars d_{QRS} for Q, R, S in $\mathcal{S}$ such that

$$C_R C_S = \sum_{Q\in\mathcal{S}} d_{QRS} C_Q.$$

The ***strata*** of this geometry are the common eigenspaces of the concurrence matrices C_S.

EXAMPLE 6 Let S be the subsets of $\mathcal{T}$ allocated to blocks in a partially balanced incomplete-block design with two associate classes, and let R be one of the two corresponding strongly regular graphs on $\mathcal{T}$. Then $\{E, S, U\}$ is a stratum geometry on $\mathcal{T}$. Clearly, this is equivalent to the association scheme $\{E, R, U \setminus R \setminus E\}$ on $\mathcal{T}$. □

EXAMPLE 7 Given a set of $r-2$ mutually orthogonal $n \times n$ Latin squares with a common transversal, let Ω consist of the $n(n-1)$ cells of the underlying $n \times n$ array which are not on the common transversal. Let S be the set-cover whose elements are the rows, columns and letters of each Latin square. Thus $k_S = n-1$ and $r_S = r$. For each α in the common transversal, let f_α be the set of elements of Ω which are in the same row, column or letter (in any of the squares) as α, and let F be the set of the f_α. Then F is a set-cover with $k_F = r(n-1)$ and $r_F = r$. It is shown in [22] that

$$C_S^2 = r(r-1)C_U - C_F + nC_S$$

$$C_S C_F = C_F C_S = r^2(r-1)C_U + (n-r)C_F$$

$$C_F^2 = r^3(r-1)C_U + r(n-r)C_F$$

and so $\{E, S, F, U\}$ is a stratum geometry. However, it is not equivalent to an association scheme unless $r = 2$ or $r = n+1$. □

Problem 9 Develop a good theory of stratum geometries.

Given a structure $(\Omega, \mathcal{R})$ with identifiable strata $(W_\theta)_{\theta\in\Theta}$ say, a design $(\Omega, \mathcal{T}, \phi)$ is defined to be ***orthogonal*** if there is an orthogonal direct-sum decomposition $\bigoplus_{\theta\in\Theta} T_\theta$ of $\mathbf{R}^{\mathcal{T}}$ such that $T_\theta\phi_* \leq W_\theta$ for θ in Θ. The design in Example 4 is orthogonal, but most incomplete-block designs are not. More generally, given an orthogonal direct-sum decomposition $\bigoplus_{\lambda\in\Lambda} T_\lambda$ of $\mathbf{R}^{\mathcal{T}}$, the design $(\Omega, \mathcal{T}, \phi)$ is *orthogonal with respect to* $\bigoplus_\theta W_\theta$ *and* $\bigoplus_\lambda T_\lambda$ if there is a function $f: \Lambda \to \Theta$ such that $T_\lambda\phi_* \leq W_{\lambda f}$ for λ in Λ. The group epimorphism designs in Section 3 are orthogonal with respect to the two group block structures. In an orthogonal design each T_λ is estimated only in its stratum $W_{\lambda f}$.

Since strata orthogonal to $\mathbf{R}^{\mathcal{T}}\phi_*$ are not used in the analysis, there is no need for them to be identifiable. To illustrate this, consider the structure on Ω defined by a transitive group G of permutations of Ω. The corresponding symmetric relations are the symmetrized orbits of G on $\Omega \times \Omega$, which have identifiable strata if and only if the permutation character π of G is real-multiplicity-free [20, 13]. If G is Abelian, the strata are the character subspaces of $\mathbf{R}^\Omega$ defined in Section 2. Even if π is not real-multiplicity-free, its decomposition into real-irreducible characters readily locates the identifiable strata.

Figure 2: Design in Example 8

	1	2	3	4	5
2		3	5	1	4
1	4		2	5	3
4	5	1		3	2
3	2	5	4		1
5	3	4	1	2	

EXAMPLE 8 Let Ω be the set of ordered pairs from an n-set, omitting self-pairs, and let G be $\mathbf{S}_n$. Let R, C, Q be the partitions of Ω into rows, columns, reverse-pairs respectively. Then the identifiable strata are shown in [5, 54] to be W_0, W_Q and W_E, where $W_0 = V_U$ (the constant vectors), $W_Q = V_Q \cap (V_R + V_C)^\perp$ and $W_E = (V_Q + V_R + V_C)^\perp$ and the V-subspaces are defined as in (4). Suppose that ϕ is a design satisfying $[E\,;E]$, $[R\,;U]$ and $[C\,;U]$. Then ϕ_* lies entirely within the identifiable strata. If, further, there is an involutory permutation τ of $\mathcal{T}$ such that $\phi\tau = \rho\phi$, where $\rho\colon\Omega \to \Omega$ is the function which is equal to $Q \setminus E$ as a subset of Ω, then ϕ is orthogonal. Put $F = E \cup \tau \subseteq \mathcal{T} \times \mathcal{T}$ and define subspaces T_0, T_F, T_E of $\mathbf{R}^{\mathcal{T}}$ by $T_0 = V_U$, $T_F = V_F \cap V_U^\perp$ and $T_E = V_F^\perp$. Then $T_0\phi* = W_0$, $T_F\phi_* \le W_Q$ and $T_E\phi_* \le W_E$. An example is shown in Figure 2 with $n = 6$, $\dim(W_Q) = 9$, $\dim(W_E) = 10$, $\dim(T_F) = 2$ and $\dim(T_E) = 2$. □

5 Incomplete-block designs

As in Examples 1 and 5, assume that $(\Omega, \mathcal{T}, \phi)$ is an incomplete-block design with a partition B of Ω into blocks. The strata are W_0, W_B and W_E. For simplicity, we assume here that ϕ is equireplicate, with replication r, and B is uniform, with blocks of size k. (The statistical literature has also developed the theory for unequal replication and unequal block sizes. The former is handled by simply applying a diagonal linear transformation to $\mathbf{R}^{\mathcal{T}}$. The latter at first sight just makes formulas more complicated. However, the matrices B and U do not commute if B is not uniform, so (E) is not satisfied. Moreover, among the many methods of justifying assumption (6) (see [81]), most do not work if blocks have different sizes.)

Let P_E and $P_{\mathcal{T}}$ be the orthogonal projectors of $\mathbf{R}^\Omega$ onto W_E and $\mathbf{R}^{\mathcal{T}}\phi_*$ respectively, and put

$$L = \phi_* P_E P_{\mathcal{T}} \phi_*^{-1}. \tag{7}$$

In fact,

$$L = E - (rk)^{-1} C_B, \tag{8}$$

where C_B is the concurrence matrix for B. When expressed as a matrix with respect to the canonical basis of $\mathbf{R}^{\mathcal{T}}$, this linear transformation L is known as the *information matrix* of the design. For orthogonal eigenvectors v and w of L, the estimators of the effects of v and w are uncorrelated. Furthermore, the variance of the estimator of the effect of v is directly proportional to the eigenvalue of $\text{Cov}(y)$ on W_E and to the squared length of v, and inversely proportional to the eigenvalue of L on v. It is thus moderately easy to interpret the results of the experiment in terms of an eigenvector basis for L.

If T_0 is the space of constant vectors in $\mathbf{R}^{\mathcal{T}}$ then $T_0\phi_* = W_0$ and hence L is zero on T_0. The eigenvalues of L on $T_0^{\perp}$ are called the *efficiency factors* of the design.

If $\mathcal{T}$ is unstructured then every efficiency factor of ϕ should be as large as possible. However, Equation (8) shows that the efficiency factors have a fixed sum. Various measures $\mathcal{E}$ of the overall efficiency of an incomplete-block design have been proposed. A design ϕ which maximizes $\mathcal{E}$ over a class Φ of designs is defined to be

1. *A-optimal* over Φ if $\mathcal{E}$ is the harmonic mean of the efficiency factors (A for *average variance*);

2. *D-optimal* over Φ if $\mathcal{E}$ is the geometric mean of the efficiency factors (D for *determinant*);

3. *E-optimal* over Φ if $\mathcal{E}$ is the minimum efficiency factor (E for *extreme*).

More generally, a design is

4. *universally optimal* over Φ if it maximizes every concave function $\mathcal{E}$ of information matrices which is invariant under permutations of $\mathcal{T}$ and which is non-decreasing in positive scalar multiples [62].

A balanced incomplete-block design is optimal with respect to all of these criteria. However, Fisher's inequality shows that $r \geq k$ for a balanced incomplete-block design, while many practical considerations force $r < k$. Statisticians need methods of constructing optimal (or nearly optimal) designs for all prescribed values of $|\mathcal{T}|$, r and k with k dividing $r|\mathcal{T}|$.

Problem 10 For any of (1)–(4), find an algorithm (with perhaps finitely many special cases) for constructing, for all values of $|\mathcal{T}|$, r and k with k dividing $r|\mathcal{T}|$, a design ϕ such that $\mathcal{E}(\phi) \geq 0.99\mathcal{E}(\phi_{\text{opt}})$, where ϕ_{opt} is optimal for $\mathcal{E}$.

Problem 10 has proved to be very hard, although certain sub-classes of design have been identified as likely to contain optimal designs.

DEFINITION An incomplete-block design $((\Omega, B), \mathcal{T}, \phi)$ is

(i) ***partially balanced*** [27] if there is an association scheme $\mathcal{R}$ on $\mathcal{T}$ such that C_B is a linear combination of the matrices R in $\mathcal{R}$—that is, ϕ satisfies $[B\,; R]$ for all R in $\mathcal{R}$;

(ii) ***a regular-graph design*** [57] if there is a regular graph R on $\mathcal{T}$ such that $C_B = rE + c(U - E) + R$ for some integer c;

(iii) ***a strongly-regular-graph design*** [36] if it is a regular-graph design for a strongly regular graph R.

Partially balanced designs have appealing algebraic properties, because the eigenspaces of L are the common eigenspaces of the Bose-Mesner algebra [26]; that is, the strata of Section 4. However, some partially balanced designs are far from optimal, and so statisticians have turned away from the class as a whole. Regular-graph designs seem intuitively to have a fairer disposition of treatments among blocks, but examples are known of regular-graph designs which are poorer than some designs which are not regular-graph designs. Nonetheless, much empirical evidence suggests that the best regular-graph designs are optimal.

CONJECTURE 5.1 (JOHN AND MITCHELL [57]) For given values of $|\mathcal{T}|$, r and k, if the class of regular-graph designs is not empty then it contains a design optimal over all incomplete-block designs for $|\mathcal{T}|$, r and k.

Problem 11 Prove Conjecture 5.1 or find a counter-example.

Interestingly, we know that Conjecture 5.1 is false if the condition of equal replication is relaxed and the fixed value of r replaced by a fixed value of $|\Omega|$. Counter-examples are in [60]. Many related conjectures are given in [59]. They all look very plausible, but I shall not be surprised if one of them turns out to be false, because optimality does not always behave plausibly. For example, if blocks can be regarded as subsets of $\mathcal{T}$ then taking the complementary set of blocks of size k does not always preserve optimality, as Example 9 shows. Nor is it known whether optimality is preserved by taking the complementary design in the sense of replacing each block by its complement in $\mathcal{T}$: see [59].

Since strongly-regular-graph designs combine the good features of both partially balanced designs and regular-graph designs, it is tempting to replace 'regular' by 'strongly regular' in Conjecture 5.1. Indeed, strongly-regular-graph designs are known to be optimal in some circumstances [32, 33, 36, 53]. However, the resulting statement is not true.

EXAMPLE 9 Let $|\mathcal{T}| = 10$ and $k = 2$. If $r = 3$ then the edges of the Petersen graph form an A-optimal design. However, for $r = 6$ the complement of this graph is not an A-optimal design. A better design consists of the edges of the graph in which all vertices of a 6-circuit are joined to all 4 vertices of a null graph. □

CONJECTURE 5.2 (CHENG [35]) If $k = 2$, $r\,|T|$ is even and $2r \geq |T| \geq r+1$, then there exists an optimal design consisting of the edges of a graph R in which all $(|T| - r)$ vertices of a null graph are joined to all vertices of a graph R' and the edges of R' form an optimal design for r treatments with replication $2r - |T|$.

Problem 12 Prove Conjecture 5.2 or find a counter-example.

Even when $k > 2$ (multi-)graphs are a useful way of describing incomplete-block designs, although a given graph no longer corresponds uniquely to a design, or, indeed, to any design at all. The *variety-concurrence graph* $\mathcal{G}_\phi$ is defined in [73] to have vertex-set T and $(t,u)C_B$ edges joining treatments t and u. Cheng [34] showed that D-optimality of ϕ is equivalent to maximizing the number of spanning trees of $\mathcal{G}_\phi$. Similarly, Paterson [68] suggests that A-optimality may be linked to the number of circuits (of various sizes) in $\mathcal{G}_\phi$. Specifically, let $c_{m,\phi}$ be the number of ordered m-circuits in $\mathcal{G}_\phi$, let $c_\phi = (c_{m,\phi})_{m\geq 3}$ and define $\prec$ on sequences by $c \prec d$ if there is an integer l such that $c_m = d_m$ for $m < l$ and $c_l < d_l$.

CONJECTURE 5.3 (PATERSON [68]) If ϕ is A-optimal and ϕ' is not A-optimal then $c_\phi \prec c_{\phi'}$.

Among designs in which all non-trivial concurrences are equal to 0 or 1 (so that $\mathcal{G}_\phi$ is a simple graph) some strongly-regular-graph designs have been shown to minimize c_ϕ [69].

Problem 13 Prove Conjecture 5.3 or find a counter-example.

6 General balance

The analysis of incomplete-block designs referred to in Section 5 uses information in stratum W_E only. From the mathematics of the model there is no reason not to use stratum W_B. For a non-orthogonal design, some vectors in $\mathbf{R}^T\phi_*$ have non-zero projection onto both W_E and W_B, and so there is information to be gained on them from both strata. The traditional reason for ignoring W_B was that the eigenvalue of $\mathrm{Cov}(y)$ on W_B was assumed to be much larger than that on W_E, but that is not always true.

The information matrix for each stratum is defined as in Equation (7). For an incomplete-block design, $L_0 + L_B + L_E = E$ and T_0 is an eigenspace of all three matrices and $T_0^{\perp} = \ker L_0$. Hence the eigenspaces of L_B are the same as those of L_E, and so estimation in W_B uses the same treatment vectors as estimation in W_E. This is obviously convenient, and also has desirable statistical properties: see [52]. However, a design with more strata may not have this property.

DEFINITION A design $(\Omega, \mathcal{T}, \phi)$ with identifiable strata $(W_\theta)_{\theta\in\Theta}$ in $\mathbf{R}^\Omega$ is *generally balanced* [65] if the information matrices $(L_\theta)_{\theta\in\Theta}$ have common eigenspaces. It is *generally balanced with respect to* a direct-sum decomposition $\bigoplus_{\lambda\in\Lambda} T_\lambda$ of $\mathbf{R}^{\mathcal{T}}$ if each T_λ is a (sub-)eigenspace of every L_θ.

EXAMPLE 10 Let $\mathcal{T}$ be the set of size $n(n-1)$ in Example 7, with set-covers S and F. Let Ω have $rn(n-1)$ plots, partitioned into r super-blocks of n blocks of size $n-1$. A design ϕ may be constructed so that the images of the blocks are the elements of S and the image of each super-block is the whole of $\mathcal{T}$. This is called a *rectangular lattice* design [50]. It is generally balanced with respect to the four strata of the stratum geometry on $\mathcal{T}$ [22]. □

Even for designs with identifiable strata, statisticians use several methods of estimating treatment effects. These agree for orthogonal designs, and diverge somewhat for non-orthogonal generally balanced designs except in the simplest cases. For example, Nelder's method [66] treats the two non-constant strata of an incomplete-block design symmetrically, while Patterson and Thompson's [72] does not. There is no agreement whatsoever for designs which are not generally balanced.

Houtman and Speed [52] defined an interesting generalization of partially balanced designs and showed that they are generally balanced.

THEOREM 6.1 Let $(\Omega, \mathcal{R})$ and $(\mathcal{T}, \mathcal{S})$ be association schemes. If the design $\phi: \Omega \to \mathcal{T}$ satisfies $[R\,;\,S]$ for all R in $\mathcal{R}$ and all S in $\mathcal{S}$ then ϕ is generally balanced.

For structures defined by partitions, the information matrices span the same algebra as the concurrence matrices of the partitions. Hence the design is generally balanced if and only if these concurrence matrices commute [20].

If $\mathcal{T}$ is structured, it is most often a factorial block structure. Then the design is said to have *factorial balance* [91] if it is generally balanced with respect to the decomposition $\bigoplus_{J\subseteq I} W_J$ in Section 3, and to have *orthogonal factorial structure* [58] if it is generally balanced with respect to a refinement of $\bigoplus_J W_J$. The latter term is slightly unfortunate, because it refers neither to *orthogonal* designs nor to *structure* on Ω or $\mathcal{T}$, but the concept is important, as [49] shows.

The other common structure on $\mathcal{T}$ is that defined by a transitive group G of permutations of $\mathcal{T}$. If the structure on Ω is defined by partitions R then ϕ is a *G-design* if, for every R, there is a monomorphism ψ_R from G into the symmetric group on Ω such that $G\psi_R$ preserves R and $(g\psi_R)\phi = \phi g$ for all g in G (there is no requirement that ψ_R be independent of R).

THEOREM 6.2 [20] Let Ω have a structure defined by partitions and let G be a transitive group of permutations of $\mathcal{T}$. If $(\Omega, \mathcal{T}, \phi)$ is a G-design and the permutation character of G on $\mathcal{T}$ is real-multiplicity-free then $(\Omega, \mathcal{T}, \phi)$ is generally balanced. In particular, if G is Abelian then $(\Omega, \mathcal{T}, \phi)$ is generally balanced with respect to the character decomposition of $\mathbf{R}^{\mathcal{T}}$ given in Section 2.

If G is Abelian, G-designs are called *generalized cyclic* designs by some authors [55, 39]. The efficiency factors for the different irreducible characters and the different strata can be easily calculated [20, 14].

Analysis of non-orthogonal generally balanced designs often involves iterative methods, and convergence can be very slow if any of the spaces $T_\lambda\phi_*$ has non-zero projection onto three or more strata (where T_λ is a common eigenspace of the information matrices). The design is therefore defined to have *adjusted orthogonality* [40, 41, 64] if there is a decomposition $\bigoplus_{\lambda\in\Lambda}$ of $\mathbf{R}^{\mathcal{T}}$ such that (i) ϕ is generally balanced with respect to this decomposition, and (ii) each $T_\lambda\phi_*$ has non-zero projection onto at most two strata. Every generally balanced design has a $\Theta \times \Lambda$ table of efficiency factors (see [52]) with non-negative entries summing to 1 in each column. The design has adjusted orthogonality if each column has at most two non-zero entries, and is orthogonal if each column has exactly one non-zero entry.

To illustrate these ideas, consider the poset block structure on Ω consisting of 2 blocks each split up into a rectangular array with n rows and m columns, where $n \geq 3$ and $m \geq 3$ (see [71, 16]). The strata are, in an obvious notation, W_0, W_B, W_R, W_C and W_E, of dimensions 1, 1, $2(n-1)$, $2(m-1)$ and $2(n-1)(m-1)$. Suppose that there are nm treatments, and the restriction of ϕ to each block is a bijection. For $i = 1, 2$, the rows and columns of the i-th block induce partitions R_i and C_i of $\mathcal{T}$ which are orthogonal in the strong sense that each class of R_i meets each class of C_i in exactly one treatment. Now adjusted orthogonality is equivalent to R_1 being orthogonal to C_2 and R_2 orthogonal to C_1, in this strong sense. Defining the V-subspaces of $\mathbf{R}^{\mathcal{T}}$ as in Equation (4), put $T_R = (V_{R_1} + V_{R_2}) \cap T_0^\perp$, $T_C = (V_{C_1} + V_{C_2}) \cap T_0^\perp$ and $T_E = (V_{R_1} + V_{R_2} + V_{C_1} + V_{C_2})^\perp$. Then $T_0\phi_* = W_0$, $T_R\phi_* \leq W_R \oplus W_E$, $T_C\phi_* \leq W_C \oplus W_E$ and $T_E\phi_* \leq W_E$.

Let ϕ_R and ϕ_C be the two quotient block designs of ϕ, obtained by ignoring the columns and rows of Ω respectively. As far as efficiency factors in W_E are concerned, ϕ is optimal if it has adjusted orthogonality and ϕ_R and ϕ_C are both optimal: see [56]. Both ϕ_R and ϕ_C are 2-replicate resolvable incomplete-block designs: as such, they can be identified with symmetric block designs $\hat{\phi}_R$ and $\hat{\phi}_C$ respectively [89]. The plot-set of $\hat{\phi}_R$ is $\mathcal{T}$ with the partition R_1; the treatments are the classes of R_2; and $\hat{\phi}_R$ is defined by

$$t\hat{\phi}_R = \text{class of } R_2 \text{ containing } t.$$

The design $\hat{\phi}_C$ is defined similarly. The identification $\hat{}$ preserves A-optimality: see [89].

If $n = m \neq 6$ then ϕ is optimal if it is a lattice square [92]; that is, R_1, R_2, C_1 and C_2 are the four partitions of an $n \times n$ Graeco-Latin square. If $n < m$ then the block size of $\hat{\phi}_C$ is less than the number of treatments of $\hat{\phi}_C$, so if $\hat{\phi}_C$ is A-optimal then it must be an incomplete-block design [38]. However, the block size of $\hat{\phi}_R$ is greater than its number of treatments: for such a design to be optimal many people conjecture [59] that the number of occurrences of any two treatments in any one block differ by at most 1.

DEFINITION For $3 \leq n < m$, a *double Youden rectangle* with n rows and m columns is a set of partitions R_1, R_2, C_1, C_2 of an nm-set T into classes of size m, m, n, n respectively such that

(i) R_1 and R_2 are both orthogonal to both C_1 and C_2;

(ii) the symmetric block design defined by C_1 and C_2 is a balanced incomplete-block design;

(iii) the symmetric block design defined by R_1 and R_2 is a balanced block design in which the number of occurrences of any two treatments in any one block differ by at most 1.

Double Youden rectangles have appeared in the literature under the guise of designs for two sets of treatments in a rectangular set of plots, illustrating yet again how a single combinatorial object may form the basis of several different designs. For $m = n+1 \notin \{4, 7\}$ Hedayat et al. [51] gave a construction for double Youden rectangles using a common transversal of an $n \times n$ Graeco-Latin square; Preece [74] completed the $n \times (n+1)$ series by exhibiting a 6×7 double Youden rectangle. Some other double Youden rectangles are known: 4×7 in [37], 7×15 in [74], 4×13 in [75].

Problem 14 Find other double Youden rectangles, preferably in infinite classes.

As usual, designs are needed even for values of n and m for which balance is impossible. I have extended Hedayat et al.'s method to $m = n + 2$, using a pair of disjoint common transversals of an $n \times n$ Graeco-Latin square. Consider the bipartite graph whose vertices are the rows and columns of the square and whose edges indicate intersection within the pair of transversals, and the similar graph for Greek and Latin letters. The construction method gives optimal designs if each graph is a single circuit, in which case the pair of transversals will be called *cyclic*. If n is prime to 6, such transversals can easily be constructed from the cyclic group $\mathbf{C}_n$.

Problem 15 For n not prime to 6 (especially for $n \leq 20$) find an $n \times n$ Graeco-Latin square with a cyclic pair of disjoint common transversals.

7 Randomization

In practice, the statistician does not choose just one design, but randomizes. Three different types of randomization were noted in [76]:

(i) ***randomization of treatments***: randomly choose an element h of a known group H of permutations of $\mathcal{T}$, and use the design ϕh, where ϕ is a fixed design;

(ii) ***randomization of plots***: randomly choose an element g of a known group G of permutations of Ω, and use the design $g\phi$, where ϕ is a fixed design;

(iii) ***random choice from a bag of designs***: randomly choose ϕ from a set Φ of suitable designs.

Of course, type (iii) encompasses both of the others. Authors disagree about the purposes of randomization and so there is no consensus about what constitutes correct randomization. However, most agree that the probability (over the randomization) that plot ω receives treatment t should be proportional to the replication of t, and type (i) randomization achieves this for equireplicate designs if H is transitive on $\mathcal{T}$, while type (ii) randomization achieves it for all designs if G is transitive on Ω.

One school of thought [43, 6] says that type (ii) randomization should be used, where G preserves the structure on Ω, and nothing more should be assumed about $\mathrm{Cov}(y)$ than that it is invariant under G. As we saw in Section 4, this means that Ω has identifiable strata if and only if the permutation character of G on Ω is real-multiplicity-free. If $(\Omega, \mathcal{R})$ is an association scheme with automorphism group G then each relation in $\mathcal{R}$ is a union of orbits of G on $\Omega \times \Omega$, and any of the following possibilities can occur:

(a) each relation in $\mathcal{R}$ is a single orbit of G on $\Omega \times \Omega$—in this case $(\Omega, \mathcal{R})$ is said to be ***2-homogeneous***, and (Ω, G) has identifiable strata, the strata being the same as those of $(\Omega, \mathcal{R})$;

(b) (Ω, G) has identifiable strata, but these form a proper refinement of the strata of $(\Omega, \mathcal{R})$;

(c) (Ω, G) does not have identifiable strata.

Examples of (a) are the poset block structures (see [18]) and Latin squares based on elementary Abelian 2-groups (see [7]). Examples of (b) and (c) are given in [12].

Problem 16 Find all 2-homogeneous association schemes.

An imminent solution to Problem 16 was promised in [3], but I have not seen a solution which includes the foregoing two examples.

An alternative interpretation of Fisher's [42, 43] ideas on randomization is given by Yates [90, 93], who uses type (iii) randomization and requires that it be *valid*. Validity is interpreted in non-statistical terms in [21] as follows. Let W be any stratum of Ω, and let P be the orthogonal projector of $\mathbf{R}^{\Omega}$ onto W. For each ϕ in Φ, let Q_ϕ be the orthogonal projector of $\mathbf{R}^{\Omega}$ onto $\mathbf{R}^{\mathcal{T}}\phi_* P$; further, let Q_Φ be the average of the $(Q_\phi)_{\phi\in\Phi}$. Then (randomization from) Φ is valid if and only if, for each stratum, Q_Φ is a scalar multiple of P.

Suppose that R and S are relations on Ω and $\mathcal{T}$ respectively. The set Φ of designs will be said to *satisfy* $[\![R\,;S]\!]$ if there is an integer $m(R,S)$ such that

$$(\alpha,\beta)\in R \quad\Rightarrow\quad |\{\phi\in\Phi:(\alpha\phi,\beta\phi)\in S\}| = m(R,S).$$

(Compare this with the definition of $[R\,;S]$ in Section 1.)

THEOREM 7.1 [21] Let $(\Omega,\mathcal{R})$ and $(\mathcal{T},\mathcal{S})$ be association schemes with strata $(W_\theta)_{\theta\in\Theta}$ and $(T_\lambda)_{\lambda\in\Lambda}$. Let Φ be a class of designs which are orthogonal with respect to $\bigoplus W_\theta$ and $\bigoplus T_\lambda$ and the *same* function $f\colon\Lambda\to\Theta$. Then Φ is valid if and only if Φ satisfies $[\![R\,;S]\!]$ for all R in $\mathcal{R}$ and all S in $\mathcal{S}$.

Problem 17 Extend Theorem 7.1 to non-orthogonal designs.

So far the only progress on Problem 17 has been limited to designs which are *totally balanced* in the sense that $T_0^\perp$ is an eigenspace of every information matrix: see [2, 19].

If $(\Omega,\mathcal{R})$ is a Latin square Δ and $((\Omega,\mathcal{R}),\mathcal{T},\phi)$ is required to be a Graeco-Latin square, then Theorem 7.1 shows that Φ is valid if $\Phi\cup\{\Delta\}$ is a complete set of mutually orthogonal Latin squares: see [76]. At present, this information is useful only if the side of Δ is a prime power.

Problem 18 For integers n which are not prime powers, find an $n\times n$ Latin square Δ and a valid set Φ of $n\times n$ Latin squares orthogonal to Δ.

Similarly, any complete set of mutually orthogonal Latin squares is a valid set of Latin-square designs: see [44]. However, valid sets exist for all sizes of square: if ϕ is any one Latin square and G is the automorphism group of the square array on Ω then $G\phi$ is valid. Restrictions on the Latin square create difficulties in randomization. For odd prime powers n, valid sets of quasi-complete $n\times n$ Latin squares are given in [8]. In spite of the work in [67, 17] constructing large numbers of quasi-complete $n\times n$ Latin squares for even values of n, no other valid sets of quasi-complete Latin squares are known.

Problem 19 For integers n which are not odd prime powers, find a valid set of quasi-complete $n \times n$ Latin squares.

Sometimes *restricted randomization* is required. This means that the statistician must avoid some designs produced by the usual randomization process, whilst sacrificing neither validity nor the known strata. If Ω is unstructured then any 2-transitive group G on Ω gives the same strata as the whole symmetric group: a suitable modification of this observation extends to all poset block structures [48, 18]. It is therefore sometimes useful to impose on Ω the structure of an irrelevant group G.

EXAMPLE 11 Let Ω be physically a 2×4 array, and let $|T| = 2$. If Ω is regarded as a rectangular poset block structure, and type (ii) randomization is used, then there are four strata, with dimensions 1, 1, 3 and 3. Such low dimensions cause problems in statistical inference. If there are not strong reasons for regarding the rows and columns of Ω as physically important, then Ω may be regarded as unstructured, with strata of dimensions 1 and 7. But now ordinary randomization can give a design in which one treatment is allocated to a whole row, or to three contiguous plots at one end of Ω. In [9], Ω is identified, purely for convenience, with the 3-dimensional affine space over $\mathsf{GF}(2)$, and a design $\phi_0 : \Omega \rightarrow T$ chosen, in such a way that randomization of ϕ_0 by the 2-transitive group $\mathrm{Aff}(3, 2)$ gives none of the undesirable designs described above. □

Problem 20 By identifying the 3×5 array Ω with the 3-dimensional projective space over $\mathsf{GF}(2)$, find a design ϕ_1 for 3 equally replicated treatments such that randomization of ϕ_1 by $\mathrm{PGL}(4, 2)$ gives no design in which one treatment is allocated to a whole row, or to four or more plots in contiguous columns; and find a design ϕ_2 for 5 equally replicated treatments such that randomization of ϕ_2 by $\mathrm{PGL}(4, 2)$ gives no design in which one treatment is allocated to a whole column, or to any other three contiguous plots.

My second example of restricted randomization is simple but occurs often in practice. Here Ω is physically an $r \times n$ array in which the columns are to be ignored, and $r < n$. There are n treatments, and the design should be a complete-block design (in the rows). By Theorem 7.1, each pair of plots in the same column must have a non-zero probability of being allocated the same treatment. However, we can require that no treatment occur more than twice in any column. Using particular mutually orthogonal Latin squares based on finite fields, a valid set of complete-block designs satisfying this requirement has been found in [11] for the case that n is a prime power.

Problem 21 For $r < n$ and n not a prime power, find a valid set of complete-row designs for an $r \times n$ array with the property that in each design no treatment occurs more than twice in any column.

8 Neighbour designs

Neighbour relations on Ω, such as the N_i in Example 2 or the successor relation in an experiment conducted over several time-periods, can enter the statistical assumptions in two quite different ways. If it is assumed that neighbouring plots are likely to behave similarly, then the N_i (or their equivalent) enter the model (6) for $\mathrm{Cov}(y)$; powers of the N_i may also be included. Typical neighbour relations which arise in practice are not uniform (because of end-plots), do not commute (as matrices) with block relations, and certainly do not satisfy condition (F). Therefore neighbour structures on Ω typically do not have identifiable strata, and it is not surprising that a large number of distinct methods of analysis has been proposed in the last decade.

The other reasonable assumption about the influence of neighbours is that the response ωy is affected not only by the treatment $\omega\phi$ on plot ω but also by the treatments $\omega N_i \phi$ on neighbours of ω. So far I have been implicitly assuming that the expectation $\mathsf{E}(y)$ of y lies in $\mathbf{R}^{\mathcal{T}}\phi_*$; that is, $\mathsf{E}(y) = \phi v_0$ for some v_0 in $\mathbf{R}^{\mathcal{T}}$. Now this assumption must be extended—at least for neighbour relations which are (partial) functions—to

$$\mathsf{E}(y) = \phi v_0 + \sum_i N_i \phi v_i \qquad (9)$$

where v_0 and v_i are in $\mathbf{R}^{\mathcal{T}}$. As in Sections 2 and 3, v_i may be restricted to lie in a proper subspace of $\mathbf{R}^{\mathcal{T}}$. So long as Ω has identifiable strata, there is no disagreement about the analysis for a given choice of model (9).

Whichever assumption is made about the effect of neighbours, optimal designs can be defined somewhat as in Section 5. Rather surprisingly, optimal designs are the same for the two assumptions: those designs which have *neighbour balance* in the sense of satisfying $[N_i; S]$ for certain relations S on $\mathcal{T}$. A good survey of neighbour-balanced designs is given in [82, Chapters 14 and 15].

However, it is not always sufficiently appreciated that the two assumptions demand different analyses of the data. It has been proved in [1] that if a certain model of type (9) is assumed and a complete-block design is used, then analysis of the data as though the neighbour effects were due to neighbouring *plots* will produce worse estimates of v_0 than the analysis which ignores neighbour relations altogether.

Model (9) leads to a natural generalization of my definition of design. Now there are N sets $\mathcal{T}_1, \ldots, \mathcal{T}_N$ of treatments, and N design maps $\phi_i : \Omega \to \mathcal{T}_i$. We assume that there are vectors v_i in $\mathbf{R}^{\mathcal{T}_i}$ such that

$$\mathsf{E}(y) = \sum \phi_i v_i.$$

Technically this fits into the framework of Section 3 with the additional assumption that the effect of W_J is zero if $|J| \geq 2$, but there is a considerable separate literature under the headings 'non-interacting treatments', 'successive experiments' and 'main-effects plans'.

References

[1] A. E. AINSLEY, G. V. DYKE & J. F. JENKYN: The relation between inter-plot interference and nearest neighbour analysis, preprint, Rothamsted, 1988.

[2] J.-M. AZAÏS: Design of experiments for studying intergenotypic competition, *J. Roy. Statist. Soc.* B, **49**, (1987), pp. 334–345.

[3] L. BABAI: On the abstract group of automorphisms, In: *Combinatorics* (ed. H. N. V. Temperley), *London Math. Soc. Lecture Notes*, **52**, Cambridge University Press, Cambridge, (1981), pp. 1–40.

[4] R. A. BAILEY: Patterns of confounding in factorial designs, *Biometrika*, **64**, (1977), pp. 597–603.

[5] R. A. BAILEY: Block strata and canonical strata in randomized experiments, unpublished report, Edinburgh, 1978.

[6] R. A. BAILEY: A unified approach to design of experiments, *J. Roy. Statist. Soc.* A, **144**, (1981), pp. 214–223.

[7] R. A. BAILEY: Latin squares with highly transitive automorphism groups, *J. Austral. Math. Soc.* A, **33**, (1982), pp. 18–22.

[8] R. A. BAILEY: Quasi-complete Latin squares: construction and randomization, *J. Roy. Statist. Soc.* B, **46**, (1984), pp. 323–334.

[9] R. A. BAILEY: Restricted randomization versus blocking, *Internat. Statist. Review*, **53**, (1985), pp. 171–182.

[10] R. A. BAILEY: Factorial design and Abelian groups, *Lin. Alg. Appl.*, **70**, (1985), pp. 349–368.

[11] R. A. BAILEY: One-way blocks in two-way layouts, *Biometrika*, **74**, (1987), pp. 27–32.

[12] R. A. BAILEY: Contribution to the discussion of "Symmetry models and hypotheses for structured data layouts" by A. P. Dawid, *J. Roy. Statist. Soc.* B, **50**, (1988), pp. 22–24.

[13] R. A. BAILEY: Automorphism groups of block structures with and without treatments, In: *Proceedings of the I.M.A. Workshop on Design Theory and Coding Theory* (ed. D. K. Ray-Chaudhuri), in press.

[14] R. A. BAILEY: Cyclic designs and factorial designs, In: *Proceedings of the R. C. Bose Symposium on Probability, Statistics and the Design of Experiments* (ed. K. Sen), in press.

[15] R. A. BAILEY, F. H. L. GILCHRIST & H. D. PATTERSON: Identification of effects and confounding patterns in factorial designs, *Biometrika*, **64**, (1977), pp. 347–354.

[16] R. A. BAILEY & H. D. PATTERSON: A note on the construction of row-and-column designs with two replicates, preprint, Edinburgh and Rothamsted, 1989.

[17] R. A. BAILEY & C. E. PRAEGER: Directed terraces for direct product groups, *Ars Comb.*, **25A**, (1988), pp. 73–76.

[18] R. A. BAILEY, C. E. PRAEGER, C. A. ROWLEY & T. P. SPEED: Generalized wreath products of permutation groups, *Proc. London Math. Soc.*, **47**, (1983), pp. 69–82.

[19] R. A. BAILEY & D. A. PREECE: Randomisation for a balanced superimposition of one Youden square on another, preprint, Rothamsted and East Malling, 1987.

[20] R. A. BAILEY & C. A. ROWLEY: General balance and treatment permutations, preprint, Rothamsted and Open University, 1986.

[21] R. A. BAILEY & C. A. ROWLEY: Valid randomization, *Proc. Roy. Soc. London* A, **410**, (1987), pp. 105–124.

[22] R. A. BAILEY & T. P. SPEED: Rectangular lattice designs: efficiency factors and analysis, *Ann. Statist.*, **14**, (1986), pp. 874–895.

[23] G. BIRKHOFF: *Lattice Theory*, American Mathematical Society, New York, (1948).

[24] R. C. BOSE: Mathematical theory of the symmetrical factorial design, *Sankhyā*, **8**, (1947), pp. 107–166.

[25] R. C. BOSE: On some connections between the design of experiments and information theory, *Bull. Internat. Statist. Inst.*, **38** (**4**), (1961), pp. 251–271.

[26] R. C. BOSE & D. M. MESNER: On linear associative algebras corresponding to association schemes of partially balanced designs, *Ann. Math. Statist.*, **30**, (1959), pp. 21–38.

[27] R. C. Bose & T. Shimamoto: Classification and analysis of partially balanced incomplete block designs with two associate classes, *J. Amer. Statist. Assoc.*, **47**, (1952), pp. 151–184.

[28] G. E. P. Box & J. S. Hunter: The 2^{k-p} fractional factorial designs. I., *Technometrics*, **3**, (1961), pp. 311–352.

[29] P. J. Cameron: Personal communication, 1983.

[30] P. J. Cameron: Some treelike objects, *Quart. J. Math.*, **38**, (1987), pp. 155–183.

[31] I. M. Chakravarti: Fractional replication in asymmetrical factorial designs and partially balanced arrays, *Sankhyā*, **17**, (1956), pp. 143–164.

[32] C.-S. Cheng: Optimality of certain asymmetrical experimental designs, *Ann. Statist.*, **6**, (1978), pp. 1239–1261.

[33] C.-S. Cheng: On the E-optimality of some block designs, *J. Roy. Statist. Soc.* B, **42**, (1980), pp. 199–204.

[34] C.-S. Cheng: Maximizing the total number of spanning trees in a graph: two related problems in graph theory and optimum design theory, *J. Comb. Theory* B, **31**, (1981), pp. 240–248.

[35] C.-S. Cheng: Personal communication, 1988.

[36] C.-S. Cheng & R. A. Bailey: Optimality of some two-associate-class partially balanced incomplete-block designs, preprint, Berkeley and Rothamsted, 1988.

[37] G. M. Clarke: Four-way balanced designs based on Youden squares with 5, 6 or 7 treatments, *Biometrics*, **23**, (1967), pp. 803–812.

[38] D. R. Cox: Contribution to the discussion of "The design and analysis of block experiments" by K. D. Tocher, *J. Roy. Statist. Soc.* B, **14**, (1952), p. 97.

[39] A. M. Dean & S. M. Lewis: A unified theory for generalized cyclic designs, *J. Statist. Plann. Inf.*, **4**, (1980), pp. 13–23.

[40] J. A. Eccleston & K. G. Russell: Connectedness and orthogonality in multi-factor designs, *Biometrika*, **62**, (1975), pp. 341–345.

[41] J. A. Eccleston & K. G. Russell: Adjusted orthogonality in nonorthogonal designs, *Biometrika*, **64**, (1977), pp. 339–345.

[42] R. A. Fisher: *Statistical Methods for Research Workers*, Oliver & Boyd, Edinburgh, (1925).

[43] R. A. FISHER: *The Design of Experiments*, Oliver & Boyd, Edinburgh, (1935).

[44] R. A. FISHER: Contribution to the discussion of "Statistical problems in agricultural experimentation" by J. Neyman, *J. Roy. Statist. Soc. Suppl.*, **2**, (1935), pp. 154–157.

[45] G. H. FREEMAN: Complete Latin squares and related experimental designs, *J. Roy. Statist. Soc.* B, **41**, (1979), pp. 253–262.

[46] A. GIOVAGNOLI: On the construction of factorial designs using abelian group theory, *Rend. Sem. Mat. Univ. Padova*, **58**, (1977), pp. 195–206.

[47] A. A. GREENFIELD: Selection of defining contrasts in two-level experiments, *Appl. Statist.*, **25**, (1976), pp. 64–67.

[48] P. M. GRUNDY & M. J. R. HEALY: Restricted randomization and quasi-Latin squares, *J. Roy. Statist. Soc.* B, **12**, (1950), pp. 286–291.

[49] S. GUPTA & R. MUKERJEE: *A Calculus for Factorial Arrangements*, Springer Verlag, New York, in press.

[50] B. HARSHBARGER: Preliminary report on the rectangular lattices, *Biometrics*, **2**, (1946), pp. 115–119.

[51] A. HEDAYAT, E. T. PARKER & W. T. FEDERER: The existence and construction of two families of designs for two successive experiments, *Biometrika*, **57**, (1970), pp. 351–355.

[52] A. M. HOUTMAN & T. P. SPEED: Balance in designed experiments with orthogonal block structure, *Ann. Statist.*, **11**, (1983), pp. 1069–1085.

[53] M. JACROUX: On the E-optimality of regular graph designs, *J. Roy. Statist. Soc.* B, **42**, (1980), pp. 205–209.

[54] A. T. JAMES: Analysis of variance determined by symmetry and combinatorial properties of zonal polynomials, In: *Statistics and Probability: Essays in Honor of C. R. Rao* (eds. G. Kallianpur, P. R. Krishnaiah & J. K. Ghosh), North-Holland, Amsterdam, (1982), pp. 329–341.

[55] J. A. JOHN: Generalized cyclic designs in factorial experiments, *Biometrika*, **60**, (1973), pp. 55–63.

[56] J. A. JOHN & J. A. ECCLESTON: Row-column α-designs, *Biometrika*, **73**, (1986), pp. 301–306.

[57] J. A. JOHN & T. MITCHELL: Optimal incomplete block designs, *J. Roy. Statist. Soc.* B, **39**, (1977), pp. 39–43.

[58] J. A. JOHN & T. M. F. SMITH: Two factor experiments in non-orthogonal designs, *J. Roy. Statist. Soc* B, **34**, (1972), pp. 401–409.

[59] J. A. JOHN & E. R. WILLIAMS: Conjectures for optimal block designs, *J. Roy. Statist. Soc.* B, **44**, (1982), pp. 221–225.

[60] B. JONES & J. A. ECCLESTON: Exchange and interchange procedures to search for optimal designs, *J. Roy. Statist. Soc.* B, **42**, (1980), pp. 238–243.

[61] O. KEMPTHORNE: Classificatory data structures and associated linear models, In: *Statistics and Probability: Essays in Honor of C. R. Rao* (eds. G. Kallianpur, P. R. Krishnaiah & J. K. Ghosh), North-Holland, Amsterdam, (1982), pp. 397–410.

[62] J. KIEFER: Construction and optimality of generalized Youden designs, In: *A Survey of Statistical Design and Linear Models* (ed. J. N. Srivastava), North-Holland, Amsterdam, (1975), pp. 333–353.

[63] A. KOBILINSKY: Confounding in relation to duality of finite Abelian groups, *Lin. Alg. Appl.*, **70**, (1985), pp. 321–347.

[64] S. M. LEWIS & A. M. DEAN: On adjusted orthogonality and general balance in row-column designs, preprint, Southampton and Columbus Ohio, 1988.

[65] J. A. NELDER: The analysis of randomized experiments with orthogonal block structure. II. Treatment structure and the general analysis of variance, *Proc. Roy. Soc. London* A, **283**, (1965), pp. 163–178.

[66] J. A. NELDER: The combination of information in generally balanced designs, *J. Roy. Statist. Soc.* B, **30**, (1968), pp. 303–311.

[67] C. K. NILRAT & C. E. PRAEGER: Complete Latin squares: terraces for groups, *Ars Comb.*, **25**, (1988), pp. 17–29.

[68] L. J. PATERSON: Circuits and efficiency in incomplete block designs, *Biometrika*, **70**, (1983), pp. 215–225.

[69] L. J. PATERSON & P. WILD: Triangles and efficiency factors, *Biometrika*, **73**, (1986), pp. 289–299.

[70] H. D. PATTERSON & R. A. BAILEY: Design keys for factorial experiments, *Appl. Statist.*, **27**, (1978), pp. 335–343.

[71] H. D. PATTERSON & D. L. ROBINSON: Row-and-column designs with two replicates, *J. Agric. Sci.*, **112**, (1989), pp. 73–77.

[72] H. D. PATTERSON & R. THOMPSON: Recovery of inter-block information when block sizes are unequal, *Biometrika*, **58**, (1971), pp. 545–554.

[73] H. D. Patterson & E. R. Williams: Some theoretical results on general block designs, *Congressum Numerantium*, **XV**, (1976), pp. 489–496.

[74] D. A. Preece: Some new balanced row-and-column designs for two non-interacting sets of treatments, *Biometrics*, **27**, (1971), pp. 426–430.

[75] D. A. Preece: Some partly cyclic 13 × 4 Youden 'squares' and a balanced arrangement for a pack of cards, *Utilitas Mathematica*, **22**, (1982), pp. 255–263.

[76] D. A. Preece, R. A. Bailey & H. D. Patterson: A randomization problem in forming designs with superimposed treatments, *Austral. J. Statist.*, **20**, (1978), pp. 111–125.

[77] C. R. Rao: Factorial arrangements derivable from combinatorial arrangements of arrays, *J. Roy. Statist. Soc. Suppl.*, **9**, (1947), pp. 128–139.

[78] N. J. A. Sloane: *A Handbook of Integer Sequences*, Academic Press, New York, (1973).

[79] T. P. Speed: What is an analysis of variance?, *Ann. Statist.*, **15**, (1987), pp. 885–910.

[80] T. P. Speed & R. A. Bailey: On a class of association schemes derived from lattices of equivalence relations, In: *Algebraic Structures and Applications* (eds. P. Schultz, C. E. Praeger & R. P. Sullivan), Marcel Dekker, New York, (1982), pp. 55–74.

[81] T. P. Speed & R. A. Bailey: Factorial dispersion models, *Internat. Statist. Review*, **55**, (1987), pp. 261–277.

[82] A. P. Street & D. J. Street: *Combinatorics of Experimental Design*, Oxford University Press, Oxford, (1987).

[83] T. N. Throckmorton: Structures of classification data, Ph. D. thesis, Iowa State University, 1961.

[84] T. Tjur: Analysis of variance models in orthogonal designs, *Internat. Statist. Review*, **52**, (1984), pp. 33–81.

[85] T. Verhoeff: An updated table of minimum-distance bounds for binary linear codes, *I. E. E. E. Trans. Inform. Theory*, **IT-33**, (1987), pp. 665–680.

[86] D. T. Voss: Single-generator generalized cyclic factorial designs as pseudo-factor designs, *Ann. Statist.*, **16**, (1988), pp. 1723–1726.

[87] D. T. Voss & A. M. Dean: A comparison of classes of single replicate factorial designs, *Ann. Statist.*, **15**, (1987), pp. 376–384.

[88] J. H. WILKINSON: *The Algebraic Eigenvalue Problem*, Oxford University Press, Oxford, (1965).

[89] E. R. WILLIAMS, H. D. PATTERSON & J. A. JOHN: Resolvable designs with two replications, *J. Roy. Statist. Soc.* B, **38**, (1976), pp. 296–301.

[90] F. YATES: The formation of Latin squares for use in field experiments, *Empire J. Exp. Agric.*, **1**, (1933), pp. 235–244.

[91] F. YATES: Complex experiments, *J. Roy. Statist. Soc. Suppl.*, **2**, (1935), pp. 181–247.

[92] F. YATES: Lattice squares, *J. Agric. Sci.*, **30**, (1940), pp. 672–687.

[93] F. YATES: Sir Ronald Fisher and the design of experiments, *Biometrics*, **20**, (1964), pp. 307–321.

Statistics Department
A. F. R. C. Institute of Arable Crops Research
Rothamsted Experimental Station
Harpenden, Herts. AL5 2JQ, U. K.

DEVELOPMENTS BASED ON RADO'S DISSERTATION "Studien zur Kombinatorik"

Walter A. Deuber, Bielefeld

I. PARTITION REGULAR MATRICES

The central theme of Rado's "Studien zur Kombinatorik" are systems of linear equations possessing a partition property, which is defined as follows: A matrix with integer valued coefficients is called ***k*-partition regular** if for every equivalence relation with k classes defined on the set of positive integers there are positive integers $x_1, \ldots, x_n$ all belonging to the same equivalence class forming a solution of the homogeneous system

$$A(x_1, \ldots, x_n)^T = 0.$$

If A is k-partition regular for all k then it is called ***partition regular***. There are simple matrices which are partition regular: For $A = (1, 0, -1)$ and $A = (1, 1, -2)$ the corresponding linear equation admits singletons as solutions. $A = (1, 1, -1)$ was known to be partition regular, as Schur [Sch 16] investigated this in the context of Fermat's theorem. $A = (1, 1, -3)$ is a first example of a matrix which is not partition regular. In order to see this write $x \in \mathbb{N}$ in base 5 expansion

$$x = a_{n_1} 5^{n_1} + a_{n_1 - 1} 5^{n_1 - 1} + \cdots + a_1 5 + a_0 \text{ with } a_i \in \{0, 1, 2, 3, 4\}, a_{n_1} \neq 0.$$

Define an equivalence relation on $\mathbb{N}$ with four classes, by putting x in class j iff $a_{i_0}(x) = j$, where i_0 is the smallest index i with $a_i \neq 0$. An easy calculation shows that the equation $x_1 + x_2 - 3x_3 = 0$ does not have a solution with all the x_i's in the same class. By a straigthforward depth first search one can see that for every equivalence relation with at most 3 classes the equation $x_1 + x_2 - 3x_3 = 0$ has a solution contained in one class, [Ra 69].

It was natural to ask for a characterization of all matrices with integral coefficients which are partition regular. Rado achieved this as follows.

Definition. Let $A = (a^1, \ldots, a^n)$ be a matrix with columns $a^j \in \mathbb{Z}^m$. A has the *columns property* iff there exists a partition of $\{1, \ldots, n\}$ into blocks $B_0, \ldots, B_{k-1}$ in such a way that

(i) $\sum_{i \in B_0} a^i = 0,$

(ii) $\sum_{i \in B_j} a^i \in$ linear span$_{\mathbb{Q}} < a^i | i \in B_0 \cup \cdots \cup B_{j-1} >$.

Thus A has the columns property iff its columns are linearly dependent in a very specific sense. Certain columns have to sum up to the zero vector. Then there are some columns (B_1) which sum up to a linear combination of columns in B_0, etc.

Theorem [Ra 33] *Let A be a finite matrix with integral coefficients. Then A is partition regular iff A has the columns property.*

The two parts of the proof of Rado's theorem differ very much in nature. That a matrix with the colums property has the partition property can be shown by combinatorial methods which will be discussed in a later section. The reverse implication is obtained by a thorough analysis of base q arithmetic (q prime) combined with a compactness argument:

For $x \in \mathbb{N}$ and prime q consider the base q-expansion. Thus $x = 3q^2 + 5q^1 + 0 \cdot q^0$ has the expansion 350. Let $l_q(x)$ be the last nonzero coefficient and $p_q(x)$ its position ($l_q(350) = 5, p_q(350) = 1$). For fixed prime q define an equivalence relation on $\mathbb{N}$ with $q-1$ classes by $x \sim_q y$ iff $l_q(x) = l_q(y)$. Assuming that A is partition regular, obtain a solution $(x_1, \ldots, x_n)$ of $Ax^T = 0$ with all the last nonzero coefficients in base q-expansion being the same, say l. Now start evaluating

$$0 = Ax^T = \sum a^i x_i$$

in base q arithmetic. The preimages p_q^{-1} define a partition $B_0, B_1, \ldots$ of $\{1, \ldots, n\}$. Moreover by performing the base q arithmetic at such position i with $p_q^{-1}(i) \neq \emptyset$ obtain relations

$$(i_q) \quad \sum_{i \in B_0} a^i \equiv 0 \quad (mod\ q)$$

$$(ii_q) \quad \sum_{i \in B_0 \cup \cdot \cup B_{j-1}} a^i \xi_{i,j} + l \cdot q^{m_j} \sum_{i \in B_j} a^i \equiv 0 \quad (mod\ q^{m_j+1})$$

for some $\xi_{i,j} \in \mathbb{Z}$ and all j.

Two observations finish the proof.

Observation 1 *For infinitely many primes q the partitions $B_0^q, B_1^q, \ldots$ of $\{1, \ldots, n\}$ coincide.*

This holds by the pigeon hole principle.

Observation 2 *If the relations (i_q) and (ii_q) hold for infinitely many primes q, then the relations (i) and (ii) of the columns property hold.*

This may be seen by assuming that $\sum_{i \in B_j} a^i \notin$ linear hull$_{\mathbb{Q}} < a^i | i \in B_0 \cup \cdots \cup B_{j-1} >$, (the linear span of the empty set should be $\{0\}$ in order to handle the case $i = 0$). Find a vector b with integral coefficients which is orthogonal to all a^i, $i \in B_0 \cup \cdots \cup B_{j-1}$, but not orthogonal to $\sum_{i \in B_j} a^i$. Then

$$\sum_{i \in B_0 \cup \cdots \cup B_{j-1}} a^i \xi_{i,j} + lq^{m_j} \sum_{i \in B_j} a^i \equiv 0 \quad (mod\ q^{m+1})$$

implies

$$l \cdot q^m(b, \sum_{i \in B_j} a^i) \equiv 0 \quad (mod\ q^{m+1}),$$

and thus

$$(b, \sum_{i \in B_j} a^i) \equiv 0 \quad (mod\ q)$$

which can hold for finitely many primes q's only. □

It should be noted that the proof shows that A is partition regular iff for some large prime q the system $Ax^T = 0$ has a solution in one class of the Rado equivalence relation which is defined by $x \sim_q y$ iff $l_q(x) = l_q(y)$. So for a given matrix A there is a least integer $k(A)$ such that if A is $k(A)$-partition regular then it is partition regular. Unfortunately Rado's proof gives an upper bound on $k(A)$ depending on the coefficients of A. Rado conjectured that $k(A)$ depends on the number of columns only. This conjecture is one of the intriguing open problems.

From the algorithmic point of view it has been shown that the decision problem whether A is partition regular is NP-complete [Graham, personal communication].

Let us now briefly discuss some examples. The Schur matrix $A = (1, 1, -1)$ clearly has the columns property. A one row matrix obviously has the columns property iff some of its nonzero coefficients sum up to zero. Extending Schur's approach

$$A_3 = \begin{pmatrix} 1 & 1 & 0 & -1 & 0 & 0 & 0 \\ 1 & 0 & 1 & 0 & -1 & 0 & 0 \\ 0 & 1 & 1 & 0 & 0 & -1 & 0 \\ 1 & 1 & 1 & 0 & 0 & 0 & -1 \end{pmatrix}$$

has the columns property. In general the matrix A_n belonging to the system

$$\sum_{i \in I} x_i = x_I \ ; \ I \subseteq \{1, \ldots, n\}, I \neq \phi$$

is partition regular. This fact did not play any particular role in Rado's work, but was rediscovered many times afterwards (e.g. by Folkman and Sanders) [GR 71, Sa 69] in the following forms:

Theorem *Let $k, n \in \mathbb{N}$. There exists a smallest $m = m(n, k) \in \mathbb{N}$ such that for every equivalence relation with k classes defined on $\{1, \ldots, m\}$ there exist n integers with all their sums without repetition in one class.*

Arnautov [Ar 70] used this theorem to show that countable rings admit a topology.

Another well known variant of this theorem states:

Theorem *Let $k, n \in \mathbb{N}$. There exists a smallest $m = m(n,k)$ such that for every equivalence relation with k classes defined on the subsets of $\{1,\ldots,m\}$ there exist n nonempty mutually disjoint sets with all their unions in one class.*

Both theorems admit infinite versions which we shall discuss later on. Another famous matrix W_n which is partition regular is given by the system of equations

$$x_\lambda = x + \lambda \cdot y \; ; \; \lambda = 0,\ldots,n-1.$$

Van der Waerden [vdW 27] proved that for every equivalence relation on $\mathbb{N}$ with finitely many classes and every n there exists an arithmetic progression $x + \lambda \cdot y = x_\lambda, \lambda = 0,\ldots,n-1$ with n terms all in one class. Van der Waerden's theorem is not quite the statement that W_n is partition regular, as it does not include that the difference y of the progression belongs to the same class as the progression itself. The observation that W_n is partition regular is due to Brauer [Br 28].

At a first glance it looks rather unnatural to investigate linear equations which should be solved by natural numbers. However it is easily seen that the same matrices with integral coefficients are partition regular for equivalence relations defined on $\mathbb{N}$ or on $\mathbb{Z}\backslash\{0\}$ or on the rationals $\mathbb{Q}\backslash\{0\}$: Let $k \in \mathbb{N}$ and assume that A is k-partition regular for equivalence relations defined on $\mathbb{Q}\backslash\{0\}$. Then by a standard compactness argument there is a finite subset X of $\mathbb{Q}\backslash\{0\}$ such for every equivalence relation on X with k classes the system $Ax^T = 0$ has a solution in one class. But then $lcm(X) \cdot X \subseteq \mathbb{Z}\backslash\{0\}$, and A is k-partition regular over $\mathbb{Z}$. By distinguishing between positive and negative integers obtain that A is $\lfloor k/2 \rfloor$-partition regular in $\mathbb{N}$. By inclusion if A is partition regular over $\mathbb{N}$ then it is partition regular over $\mathbb{Q}\backslash\{0\}$.

These arguments are elementary except for the one place, where a compactness argument is used. In fact Rado took much care to work out the details. He established a "Gleichmäßigkeitssatz" and did so again in the second part of his work in a more general context. At this point – at least it seems to me – Rado established a bridge between finite combinatorics and infinite combinatorics which proved to be so fruitful in many parts of his work. Moreover the compactness argument is not constructive. Thus it is the crucial place where independence results like those of Paris and Harrington [PH 77] come in.

Closing this section we report on some further developments. Rado not only considered homogeneous linear equations but also inhomogeneous systems.

Theorem [Ra 33] *The system $Ax^T = b$ is partition regular over $\mathbb{N}$ iff it admits a singleton solution $x = (x,\ldots,x)$ for some $x \in \mathbb{Z}$ and if x has to be negative or zero then A has the columns property.*

The case of real and complex matrices was also investigated by Rado.

Theorem [Ra 43] *Let A be a matrix with complex entries. Then A is partition regular over $\mathbb{C}\backslash\{0\}$ iff it is partition regular over $R\backslash\{0\}$, where R is the ring generated by the coefficients of A.*

This implies that complex partition regular matrices are characterized by appropriate columns properties [Ra 43].

Real partition regular matrices showed up later on in the framework of Euclidean Ramsey theory [EGMRSS 73,75,75]. The Euclidean Ramsey theory problem is to characterize point sets S in Euclidean space with the following property: For every k there exists an n such that for every equivalence relation with k classes defined on $\mathbf{R}^n$ there exists a congruent copy $\bar{S}$ of S contained in one class of the equivalence relation. Such sets are called *Ramsey sets.* Partition regular matrices play a crucial role in the proof of

Theorem [EGMRSS 73] *If S is Ramsey, then it is embeddable into a sphere.*

The most recent result in this area is due to Frankl and Rödl who proved

Theorem [FR 87] *Simplicial sets are Ramsey.*

A q-analog of partition regular systems can be established for Abelian groups. Let G be an Abelian group, considered as a $\mathbf{Z}$-module, and let A be a matrix with integral coefficients. Call A to be *partition regular over G* iff for every equivalence relation with finitely many classes defined on $G\backslash\{0\}$ at least one class contains a solution of the homogeneous system $Ax^T = 0$. A matrix A has the *p-columns property* (p prime) iff it has the columns property with all linear combinations taken modulo p.

Theorem [De 75] *Let G be an Abelian group and A a finite matrix with integral coefficients. A is partition regular over G iff one of the following condition holds:*

(i) $Ax^T = 0$ has a solution in $G\backslash\{0\}$ with $x_1 = x_2 = \cdots = x_n$.

(ii) For some prime p the group G contains the infinite direct sum of the cyclic groups $\mathbf{Z}_p$ and A satisfies the p-columns property.

(iii) G contains elements of arbitrarily high order or an element of infinite order and A satisfies the columns property.

The problem of non-linear partition regular systems of equations seems to be much harder. Not too much is known, in particular it is not known whether $x^2 + y^2 = z^2$ is partition regular over $\mathbf{N}$. However, this equation is partition regular over $\mathbf{R}$. A recent result on non-linear partition regular systems of equation was obtained by Lefmann:

Theorem [Lef 88] *Let A, B be integral matrices and $k, l \in \mathbf{N}$. Then the system*

$$A(x_1^{1/k}, \ldots, x_n^{1/k})^T + B(x_{n+1}^{-1/l}, \ldots, x_m^{-1/l})^T = 0$$

is partition regular over $\mathbf{N}$ iff either A and B have the columns property or there is a solution with $x_1 = \cdots = x_m \in \mathbf{N}$.

II. (m,p,c)-SETS

With the columns property of a matrix A Rado had a characterization of the partition regular matrices which was very satisfactory from the point of view of linear algebra. However it did not describe the structure of the solutions of systems of equations $Ax^T = 0$ for partition regular A. From the point of view of combinatorics it is desirable to have such a characterization. A first step in this direction was done by Rado [Ra 43] indicating, that the solutions have a certain matrix structure. It seems that in combinatorics the transformation from matrix theory to the combinatorics of sets has often been fruitful. Remember that P. Hall's famous criterion for the marriage problem in bipartite graphs presumably was hidden in matrix theoretical investigations of Frobenius.

Definition: Let m, p, c be natural numbers. A set $D \subseteq \mathbb{N}$ is an (m, p, c)-set if there exist positive integers $d_0, \ldots, d_m$ such that D consists of all numbers in the following "defining list"

$$(\star) \quad \begin{array}{rrrrr} cd_0 + l_1d_1 + & l_2d_2 + & \ldots & + l_md_m \\ cd_1 + & l_2d_2 + & \ldots & + l_md_m \\ & cd_2 + & \ldots & + l_md_m \\ & & & \vdots \\ & & & cd_m \end{array}$$

where $l_i \in \{0, \pm 1, \cdots \pm p\}$.

At a first glance these (m, p, c)-sets look rather clumsy. However the q-analog of them is well known and we encounter the rare situation, where the q-analog was known before the classical combinatorial situation.

Consider $GF(q)^{m+1}$ with canonical basis $d_0, \ldots, d_m$. Then the points of the m-dimensional projective space ($= 1$ dimensional subspaces) of $GF(q)^{m+1}$ may be described by representatives contained in the following list

$$\begin{array}{rrrrr} d_0 + l_1d_1 + & l_2d_2 + & \ldots & + l_md_m \\ d_1 + & l_2d_2 + & \ldots & + l_md_m \\ & d_2 + & \ldots & + l_md_m \\ & & & \vdots \\ & & & d_m \end{array}$$

where $l_i \in GF(q)$.

Thus (m, p, c)-sets have projective spaces over finite fields as q-analogs and therefore are the combinatorial–numbertheoretic counterpart of those. But independently from this analogy they can be interpreted in different ways.

— As $a + \lambda b, \lambda = 0, \ldots, n$ is an arithmetic progression, it makes sense to consider each row

$$cd_i + l_{i+1}d_{i+1} + \cdots + l_md_m$$

of the defining list $(\star)$ as an $(m - i)$-fold arithmetic progression. And therefore (m, p, c)-sets are amalgamated systems of multiple arithmetic progressions.

— Let D be the (m, p, c)-set generated by $d_0, \ldots, d_m$ and D' the $(m-1, p, c)$-subset generated by $d_0, \ldots, d_{m-1}$. Then

$$D = \{x + l_md_m \mid x \in D', l_m \in \{0, \pm 1, \cdots \pm p\}\} \cup \{c \cdot d_m\}.$$

This recursive definition shows that D is obtained from D' by attaching to each $x \in D'$ an arithmetic progression, plus the extra element cd_m. Calling the arithmetic progression $\{x + l_m d_m | l_m \in \{0, \pm 1, \cdots \pm p\}\}$ an arithmetic neighborhood, one can view (m,p,c)-sets as iterated systems of arithmetic neighborhoods.

— Once the generators $d_0, \ldots, d_m$ of an (m,p,c)-set D are given, the elements in each line of the defining list $(\star)$ are given by their coordinates $(l_{i+1}, \ldots, l_m)$. Thus each line is a combinatorial "Hales–Jewett cube", or a "Graham–Rothschild parameter set" (cf. section III). This connection was first observed by K. Leeb [Le 75]. It renders the proofs of partition theorems for (m,p,c)-sets more lucid than the original ones.

The connection between the (m,p,c)-sets and partition-regular matrices is the following.

Observation 1 *Let (m,p,c) be given. Then there exists a partition-regular matrix A such that every solution $\{x_1, \ldots, x_n\}$ of $Ax^T = 0$ contains an (m,p,c)-set.*

In order to see this proceed by induction on m and consider the linear equations $x_i + l \cdot x_m/c = x_{i,l}$, where $x_i \in D', l \in \{0, \pm 1 \cdots \pm p\}$ occuring in the recursive definition of (m,p,c)-sets.

Observation 2 *Let A be a matrix which has the columns property. Then there exists (m,p,c) such that every (m,p,c)-set contains a solution $\{x_1, \ldots, x_n\}$ of $Ax^T = 0$.*

In order to see this it suffices to take m as the number of blocks occuring in the partition of $\{1, \ldots, n\}$ defining the columns property. c is the least common denominator of all rational coeffcients involved in the linear combinations occuring in the columns property, whereas $p = c \cdot max\{|\xi| \;\; ; \;\; \xi$ a rational coefficient in a linear combination in the columns property of $A\}$.

The crucial link is the following partition theorem for (m,p,c)-sets, which we will prove in a later section.

Theorem [De 73] *For every triple (m,p,c) of positive integers and $k \in \mathbb{N}$ there exists a triple (n,q,d) with the following property. Let $\sim$ be an equivalence relation with k classes defined on an (n,q,d)-set. Then one class of $\sim$ contains an (m,p,c)-set.*

The combination of this theorem with Observation 2 shows that matrices having the columns property are partition regular and so completes a proof of Rado's characterization of partition regular matrices as indicated in section I.

Rado observed that partition regular matrices have a direct sum property: If A, B are partition regular then so is $\begin{pmatrix} A & O \\ O & B \end{pmatrix}$. From this it is easy to conclude by an indirect argument that the following observation holds.

Observation 3 *Let $\sim$ be an equivalence relation on $\mathbb{N}$ with finitely many classes. Then one of the classes contains a solution for every partition regular system of equations $Ax^T = 0$.*

This observation contains a change of quantifiers from $\forall\exists$ to $\exists\forall$: By definition, for a given equivalence relation $\sim$ on $\mathbb{N}$ with finitely many classes, for each partition regular A there exists one class of $\sim$ containing a solution of the associated system of equations. One class may be chosen to contain a solution for all partition regular systems $Ax^T = 0$.

Observation 3 led Rado to a question, which in todays terminology may be stated as follows.

Definition Call a subset X of $\mathbb{N}$ *partition regular* iff it contains a solution of $Ax^T = 0$ for every partition regular matrix A.

Rado wondered whether in the last observation one could replace the set of all positive integers by any partition regular set. In other words, he conjectured that the property of a set to be partition regular is hereditary for partitions (equivalence relations) with finitely many classes. It is remarkable that with the matrix theoretical characterization of solutions of partition regular systems of equations he was pretty close in [Ra 43].

Theorem [De 73] *Let $X \subseteq \mathbb{N}$ be partition regular, and let $\sim$ be an equivalence relation with finitely many classes defined on X . Then at least one of the classes is partition regular again.*

Proof: By the cofinality of the class of solutions of partition regular systems of equations and the class of (m,p,c)-sets a set X is partition regular iff for every (m,p,c) it contains an (m,p,c)-set. By the partition theorem for (m,p,c)-sets one concludes that for every given (m,p,c) at least one class of an equivalence relation $\sim$ with finitely many classes defined on X contains an (m,p,c)-set. As the inclusion on (m,p,c)-sets is a directed partial order(upper bounds exist) one can choose a class of $\sim$ containing (m,p,c)-sets for all triples (m,p,c). Again by cofinality this class is partition regular. □

An obvious and classical way to generalize partition regular matrices is to consider infinite systems of equations. One way to do so would be to ask whether and how (m,p,c)-sets could be generalized in such a way that still a partition theorem holds. Obviously c should be finite. With respect to p it is well known that in van der Waerden's theorem one can not expect to have infinite progressions in one class of an equivalence relation: In order to have an equivalence relations $\sim$ with 2 classes on $\mathbb{N}$ which admits no infinite arithmetic progression contained in any class, enumerate the denumerably many progressions and in the i-th step of the construction of $\sim$ make sure that the i-th progression is not contained in one class. A more constructive proof may be obtained by putting intervals $[2^i, 2^{i+1}[$ into alternating classes. So for fixed p, c one could try to take $m = \omega$. In [CH 75, CEHR 76] the q-analog of this question is investigated and it is shown that $GF(2)$ is the only case, where positive results could be expected.

Sanders, Graham and Rothschild conjectured that the finite sum theorem and the finite union theorem mentioned above have an infinite analog. The q-analog of this is the projective case over $GF(2)$:

Theorem [Hin 74] *Let $\sim$ be an equivalence relation with finitely many classes defined on $\mathbb{N}$. Then there is an infinite collection $(x_i)_{i\in\omega}$ of natural* *numbers such that all the sums* $\sum_I x_i$ *for finite* $I \subset \mathbb{N}, I \neq \emptyset$ *are contained in one class of* $\sim$.

Theorem [Hin 74] *Let* $\sim$ *be an equivalence relation with finitely many classes defined on the set of all finite nonempty subsets of* $\mathbb{N}$. *Then there is an infinite collection* $(X_i)_{i\in\omega}$ *of pairwise disjoint finite nonempty subsets of* $\mathbb{N}$ *such that all the unions* $\cup_I X_i$ *for finite* $I \subset \mathbb{N}, I \neq \emptyset$ *are contained in one class of* $\sim$.

This is the infinite version of Rado's sum and union theorem. After the first proof by Hindman, Baumgartner [Ba 74] and Glazer gave short proofs. The most elegant way to look at it is Galvin's construction with ultrafilters. In the framework of the Stone-Čech compactification $\beta\mathbb{N}$ of $\mathbb{N}$ it boils down to the fact that a compact topological semigroup has an idempotent.

Fürstenberg and Weiss [FW 78] showed, that for every equivalence relation on $\mathbb{N}$ with finitely many classes one of these classes contains a solution for every finite partition regular system of equations and an infinite collection of numbers x_i together with all their finite sums. Applying Galvin's method in [DH 87] so far the biggest partition regular system was exhibited.

Theorem [DH 87] *For every equivalence relation on* $\mathbb{N}$ *with finitely many classes one of the classes contains a family of pairwise disjoint* (m,p,c)*-sets* $X_{m,p,c}$, *one for each* $(m,p,c) \in \mathbb{N}^3$ *and moreover all the finite sums* $\sum x_i$, *where the* x_i *are elements of the* $X_{m,p,c}$*'s, no two* x_i *belonging to the same* $X_{m,p,c}$.

We do not know how far away this is from a complete characterization of partition regular systems of equations with denumerably many finite equations.

III. COMBINATORIAL LINES AND PARAMETER SETS

In his "Studien zur Kombinatorik" Rado relied on and gave a new proof of van der Waerden's theorem on arithmetic progressions. Since then a remarkable number of new or short combinatorial proofs were given. Two aspects deserve special attention.

The first aspect is the transfer of the notion of arithmetic progression to the purely combinatorial notion of a combinatorial line or a parameter set. This was obtained by stripping arithmetic progressions of their number theoretic trappings by Hales and Jewett [HJ 63], generalized by Graham and Rothschild [GR 71] and later on rephrased in a categorical framework in the paper by Graham, Leeb and Rothschild [GLR 72].

The second aspect is that the proofs relied on a double induction. This always yielded a non primitive recursive upper bound for the van der Waerden function. Shelah 1988 succeeded in showing that the van der Waerden function is primitive recursive.

Definition Let A be a finite set (alphabet). A combinatorial line $L \subseteq A^n$ is a set of cardinality $|A|$ such that there exists a nonempty subset $I \subseteq$

$\{0,\ldots,n-1\}$ of so-called "active" coordinates and for every non-active $i \in \{0,\ldots,n-1\} \setminus I$ there exists $a_i \in A$ such that

$$L = \{(x_0,\ldots,x_{n-1}) | x_i = x_j \text{ for all } i,j \in I \text{ and } x_i = a_i \text{ for } i \notin I\}.$$

Combinatorial lines were introduced by Hales and Jewett [HJ 63].

Again the q-analog of a combinatorial line is quite familiar. By taking $A = GF(q)$ every combinatorial line in $GF(q)^n$ is an affine line. Notice however that the set $\{(ta,tb) | t \in GF(q)\}$ is an affine line in $GF(q)^2$ which is not a combinatorial line except for $a = b$.

Theorem [HJ 63] *For every finite alphabet A and positive integer r there exists a least $HJ(A,r) \in \mathbb{N}$ such that for every equivalence relation on $A^{HJ(A,r)}$ with r classes there exists a combinatorial line contained in one of the classes.*

As a corollary we can prove van der Waerden's theorem on arithmetic progressions. Let $W(k,r)$ be the smallest $w \in \mathbb{N}$ such that for every equivalence relation on $\{0,\ldots,w-1\}$ with r classes at least one of the classes contains an arithmetic progression with k terms. We claim that

$$W(k,r) \le (k-1)HJ(A,r)+1 \quad \text{where } A = \{0,\ldots,k-1\}.$$

In order to see this let $n = HJ(A,r)$ and let $\sim$ be an equivalence relation with r classes defined on $\{0,\ldots,n(k-1)\}$. Define an equivalence relation $\sim^\star$ on A^n by $(x_0,\ldots,x_{n-1}) \sim^\star (y_0,\ldots y_{n-1})$ iff $\sum x_i \sim \sum y_i$ where obviously $\sum x_i \le n(k-1)$. Then any combinatorial line contained in one class of the equivalence relation $\sim^\star$ corresponds to an arithmetic progression P. This progression has k terms and its difference is the number of active coordinates. Moreover P is contained in one class of $\sim$. □

Theorem [Sh 88] *$HJ(A,r)$ is primitive recursive.*

Shelah succeeded in proving this by giving an essentially new proof of van der Waerden's theorem. The crucial idea is in the following innocent looking lemma.

Notation Let X be a set, $l \in \mathbb{N}$. By $[X]^l$ denote the set of l-element subsets of X. We identify n and $\{0,\ldots,n-1\}$ where convenient. $A \times B$ is the cartesian product. As usual an equivalence relation on a set is identified with the mapping (partition) which assigns to each element the index of its class.

Lemma [Sh 88] *For every $m,r \in \mathbb{N}$ there exists a least $n = Sh(m,r) \in \mathbb{N}$ with the following property: Let $(\Delta_i)_{i=0,\ldots,m-1}$ be an m-sequence of mappings with*

$$\Delta_i : (\underbrace{[n]^2 \times \cdots \times [n]^2}_{i \text{ times}} \times [n]^1 \times \underbrace{[n]^2 \times \cdots \times [n]^2}_{m-i-1 \text{ times}}) \to \{0,\ldots,r-1\}.$$

Then there exist $x_i < y_i < n$ $(i = 0,\ldots,m-1)$ such that for $i = 0,\ldots,m-1$ the following equalities hold:

$$\begin{aligned}&\Delta_i(\{x_0,y_0\},\ldots,\{x_{i-1},y_{i-1}\},\{x_i\},\{x_{i+1},y_{i+1}\},\ldots,\{x_{m-1},y_{m-1}\})\\ =&\Delta_i(\{x_0,y_0\},\ldots,\{x_{i-1},y_{i-1}\},\{y_i\},\{x_{i+1},y_{i+1}\},\ldots,\{x_{m-1},y_{m-1}\}).\end{aligned}$$

Proof The proof is a simple induction on m using a standard product argument. It provides a primitive recursive upper bound, viz.

$$Sh(1,r) = r+1 \text{ and } Sh(m+1,r) \leqslant 1 + r^{(Sh(m,r))^{2m}}$$

This may be done as follows. For $m = 1$ the pigeon hole principle may be applied. Assume that the lemma holds for some $m \geqslant 1$ and arbitrary r. So consider $m+1$ and a fixed r. Let $n = 1 + r^{\binom{Sh(m,r)}{2}^m}$ and let $(\Delta_i)_{i=0,\ldots,m}$ be an $(m+1)$-sequence of mappings with

$$\Delta_i : \underbrace{[n]^2 \times \cdots \times [n]^2}_{i \text{ times}} \times [n]^1 \times \underbrace{[n]^2 \times \cdots \times [n]^2}_{m-i \text{ times}} \to \{0,\ldots,r-1\}$$

Focus attention on Δ_m and define a mapping Δ^* on $[n]^1$ with at most $r^{\binom{Sh(m,r)}{2}^m}$ values by

$$\Delta^*\{x\} = \Delta_m \left| \left(\left(\prod_0^{m-1} [Sh(m,r)]^2 \right) \times \{x\} \right) \right.$$

By the pigeon hole principle obtain $x_m < y_m < n$ with $\Delta^*(x_m) = \Delta^*(y_m)$. Thus at least Δ_m has the desired properties with respect to x_m, y_m. In order to obtain the remaining x_i's and y_i's apply the induction hypothesis to the sequence $(\Delta_i')_{i=0,\ldots,m-1}$ defined by

$$\Delta_i' = \Delta_i \left| \underbrace{[Sh(m,r)]^2 \times \cdots \times [Sh(m,r)]^2}_{i \text{ times}} \times [Sh(m,r)]^1 \times \right.$$
$$\times \underbrace{[Sh(m,r)]^2 \times \cdots \times [Sh(m,r)]^2}_{m-i-1 \text{ times}} \times \{x_m, y_m\}.$$

□

We need the following constructions:

Let B be an alphabet, and choose two fixed letters $a, b \in B$. For $x < y < n$ consider the following line in B^n

$$L_{x,y} = \{(\underbrace{a,\ldots,a}_{x \text{ times}}, \underbrace{\lambda,\ldots,\lambda}_{y-x \text{ times}}, \underbrace{b,\ldots,b}_{n-y \text{ times}}) | \lambda \in B\}.$$

Furthermore consider the point

$$P_x = (\underbrace{a,\ldots,a}_{x \text{ times}}, b,\ldots,b) \in B^n.$$

Now we are ready to give Shelah's proof of Hales-Jewett's theorem. Proceed by induction on the size of the alphabet A.

Assume that $HJ(A,r)$ exists and is primitive recursive for some $A \neq \emptyset$. Let $a \in A$ be some fixed element, $b \notin A$ and $B = A \cup \{b\}$. Let

$$m = HJ(A,r)$$

$$n = Sh(m, r^{(|A|+1)^{m-1}}).$$

We claim that $HJ(B,r) = HJ(A \cup \{b\},r) \leq n \cdot m$. This implies that HJ is a primitive recursive function. In order to see the claim let

$$\Delta : (B^n)^m \to \{0,\ldots,r-1\}$$

be given. In order to apply the lemma define a sequence of mappings Δ_i where $i = 0,\ldots,m-1$ with Δ_i defined on $\underbrace{[n]^2 \times \cdots \times [n]^2}_{i \text{ times}} \times [n]^1 \times \underbrace{[n]^2 \times \cdots \times [n]^2}_{m-i-1 \text{ times}}$. To do so choose

$$0 \leq x_i < y_i < n \text{ for } i = 0,\ldots,m-1.$$

Consider $C_{x_i} = \prod_{j=0}^{i-1} L_{x_j,y_j} \times P_{x_i} \times \prod_{j=i+1}^{m-1} L_{x_j,y_j}$, which is an $(m-1)$-dimensional subcube of B^{nm} and put

$$\Delta_i(\{x_0,y_0\},\ldots,\{x_i\},\ldots,\{x_{m-1},y_{m-1}\}) = \Delta|C_{x_i}.$$

Thus Δ_i is a coloring with at most $r^{(|A|+1)^{m-1}}$ colors. By Shelah's lemma there exist $x_i^0 < y_i^0 < n$ for $i = 0,\ldots,m-1$ such that

$$\Delta|C_{x_i^0} = \Delta|C_{y_i^0}.$$

Consider next the m-dimensional cube

$$C = \prod_{j=0}^{m-1} L_{x_j^0,y_j^0}.$$

The construction yields that for $z \in C$ every single active coordinate z_k which equals b may be replaced by the fixed $a \in A$ without changing colors. By transitivity this means that the active coordinates in C may avoid b at all, without changing colors. Thus within C we have an m-dimensional subcube C^* with all its active coordinates over A. Applying the induction hypothesis to C^* obtain a monochromatic line L^* over the alphabet A. But changing its active coordinates back into b, by construction, does not change the color. So the extended line L over the alphabet B is monochromatic too.

Shelah's proof yields a sharper result. Define the Ackermann hierarchy by $f_1(i) = i+1, f_2(i) = 2i$ and $f_{n+1}(i) = f_n o \ldots o f_n(1)$ where the iteration is i-fold. Then $HJ(A,t) \leq f_4(c \cdot |A| + t)$ for some constant c. Is this best possible?

In the proof of Shelah's result we skipped one step by saying that a standard product argument has to be used. Such an argument occured implicitely in van der Waerden's proof. Rado used it again and again in his "Studien zur Kombinatorik" and in the second part he formulated it in a general setting. In an oversimplified way one could say Ramsey-partition theory nowadays is the artful iteration of the pigeonhole principle and product arguments.

In its simplest form the product argument can be stated as follows.

Definition Let X be a set and $\mathcal{X}$ a collection of subsets of X. $(X, \mathcal{X})$ is *k-partition regular*, iff for every equivalence relation on X with k classes at least one class contains a member of $\mathcal{X}$.

Product argument 1st part: *Let X be a set and $(Y, \mathcal{Y})$ be $k^{|X|}$-partition regular. Then for every equivalence relation on $X \times Y$ with k-classes there exists a $\tilde{Y} \in \mathcal{Y}$ such that for every $y, y' \in \tilde{Y}$ and $x \in X$ the pairs (x, y) and (x, y') belong to the same class.*

Product argument 2nd part: *Let $(X, \mathcal{X})$ be k-partition regular and $(Y, \mathcal{Y})$ be $k^{|X|}$-partition regular, then the cartesian product*

$$\{\tilde{X} \times \tilde{Y} | \tilde{X} \in \mathcal{X}, \tilde{Y} \in \mathcal{Y}\}$$

is k-partition regular.

Observe that the first part may be proved by considering for each $y \in Y$ the restriction of the equivalence relation to $\{y\} \times X$ and applying the $k^{|X|}$-partition regularity of $\mathcal{Y}$. The second part then follows by applying the k-partition regularity of $\mathcal{X}$.

By generalizing combinatorial lines one can consider parameter sets.

Definition Let A be a finite alphabet. An *m-parameter set* $P \subseteq A^n$ is a set of cardinality $|A|^m$ such that there exist m pairwise disjoint nonempty subsets $I_1, \ldots, I_m \subseteq \{0, \ldots, n-1\}$ and for every $i \in \{0, \ldots, n-1\} \backslash (I_1 \cup \cdots \cup I_m)$ there exists $a_i \in A$ such that $P = \{(x_0 \ldots x_{n-1}) | \; x_i = x_j$ for all $i, j \in I_k, k = 1, \ldots, m$ and $x_i = a_i$ for $i \notin I_1 \cup \cdots \cup I_m\}$.

Corollary [GR 71] *Let A be a finite alphabet. For every $m, r \in \mathbb{N}$ there exists a smallest $n(A, m, r) \in \mathbb{N}$ such that for every equivalence relation with r classes defined on A^n one of the classes contains an m-parameter set.*

For a proof it suffices to apply the Hales-Jewett Theorem with the alphabet A^m.

Here we resist the temptation to report on the beautiful Ramsey type results on parameter sets, finite vector spaces, graphs and boolean algebras etc, which are based on Hales-Jewett's theorem. For a recent account see [PV 85a]. These results are important and Rado has his personal share in the development of Ramsey theory. On the other hand, Ramsey's theorem is not mentioned in the "Studien zur Kombinatorik". It seems that most mathematicians took note of Ramsey's theorem only after the paper by Erdös and Szekeres 1935 [ES 35].

Now we are ready to give a simple proof of the partition theorem for (m, p, c)-sets. Let us concentrate on $c = 1$. Let m, p and k be given, n be large enough and $\sim$ an equivalence relation with k classes defined on an $(n,p,1)$-set D. The first row of the defining scheme $(\star)$ is

$$R_1 = \{d_0 + l_1 d_1 + \cdots + l_n d_n \quad ; \quad l_i \in \{0, \pm 1, \cdots \pm p\}\} .$$

Considering $A = \{0, \pm 1, \cdots \pm p\}$ observe that R_1 is isomorphic to A^n and thus $\sim$ is defined on A^n. By the partition theorem for parameter sets one equivalence class contains an n_1-parameter set D_1 (n_1 should still be large enough). Going back to R_1 obtain

$$R_1' = \{d_0' + l_1 d_1' + \ldots l_{n_1} d_{n_1}' ; l_i \in A\} .$$

contained in one class of $\sim$. Iterating over all rows of D obtain an $(n^\star, p, 1)$-set $D^\star \subset D$ such that each row of $D^\star$ is contained in one class of $\sim$. By the pigeon hole principle $m+1$ rows are contained in the same class of $\sim$. Thus there is an $(m,p,1)$-set contained in one class of $\sim$.

For $c \neq 1$ proceed analogously taking care that the leading coefficients are multiples of c.

IV. GRAPHS WITH ARITHMETIC STRUCTURE

It is not surprising that ideas originating from graph theory found their way into partition theory. A ***set system*** $\mathcal{E}$ or a ***hypergraph*** on a set V is a family of subsets of V which are called "edges". $\mathcal{E}$ is ***uniform*** if all edges have the same cardinality. The ***chromatic number*** $\chi(\mathcal{E})$ is the least positive integer r such that there exists an r-coloring $\Delta : V \to \{0, \ldots, r-1\}$ without any monochromatic edge. Thus $r-1$ is the maximum integer k such that for every equivalence relation on V with k classes at least one class contains an edge of $\mathcal{E}$. In this way chromatic numbers and partition theorems are closely related.

One appealing problem in graph theory was to determine how sparse graphs with given chromatic number could be.

Definition A *cycle* in a hypergraph $\mathcal{E}$ is given by a sequence $x_0, \ldots, x_{g-1}$ of pairwise distinct vertices together with a sequence $E_0, \ldots, E_{g-1}$ of pairwise distinct edges such that $\{x_i, x_{i+1}\} \subseteq E_i$ for $i < g-1$ and $\{x_{g-1}, x_0\} \subseteq E_{g-1}$. The ***girth*** of $\mathcal{E}$ is the length g of a shortest cycle.

The following theorem has a long history.

Theorem [EH 66,Lo 68,NR 79] *Let c, g, m be positive integers. There exist m-uniform hypergraphs $\mathcal{E}$ with $\chi(\mathcal{E}) > c$ and girth $(\mathcal{E}) > g$.*

Now we consider hypergraphs with arithmetic structure.

Definition Let $V \subseteq \mathbb{N}$. The edges of the ***k-van der Waerden hypergraph*** on V are the arithmetic progressions with k terms contained in V. A subset E of V is an edge of the ***m-Rado hypergraph*** on V iff there exists an m element set $\{x_1, \ldots, x_m\} \subset V$ such that $E = \{\sum_I x_i | I \subset \{1, \ldots, m\}, I \neq \emptyset\}$.

Van der Waerden's theorem on arithmetic progressions states that for each r and k there exists an n such that the chromatic number of the k-van der Waerden hypergraph on $\{0, \ldots, n-1\}$ exceeds r. But certainly this hypergraph contains a lot of short cycles, even cycles of length 2. This fact is used in the standard proofs of van der Waerden's theorem. Spencer asked whether such cycles are really necessary. He proved by probabilistic methods:

Theorem [Sp 75] *Let g, k, r be positive integers. There exists an n and a subgraph $\mathcal{E}$ of the k- van der Waerden hypergraph on $\{0, \ldots, n-1\}$ with girth $(\mathcal{E}) > g$ and $\chi(\mathcal{E}) > r$.*

Observe that this theorem does not state that $\mathcal{E}$ is the van der Waerden hypergraph on some set $V \subset \{0, \ldots, n-1\}$. Moreover the probabilistic method does not prove this, as $\mathcal{E}$ covers almost all $i < n$. Spencer [Sp 75] conjectured the following theorem:

Theorem [PV 88] *Let g, k, r be positive integers. Then there exists a set $V \subset \mathbb{N}$ such that the k-van der Waerden hypergraph on V has girth $(\mathcal{E}) > g$ and $\chi(\mathcal{E}) > r$.*

An analogous theorem holds for Rado hypergraphs:

Theorem [PV 88] *Let g, m, r be positive integers. Then there exists a set $V \subseteq \mathbb{N}$ such that the m-Rado hypergraph $(V, \mathcal{E})$ has girth $(\mathcal{E}) > g$ and $\chi(\mathcal{E}) > r$.*

The case $m = 2$ of this theorem was proved by Nešetřil and Rödl [NR 86], who raised the question whether the above theorem holds.

Also with respect to Hales-Jewett's theorem a sparse version has been proved in [PV 88]. This improves a previous result of Rödl [Rö 81] which was proved by probabilistic methods. This sparse version of Hales-Jewett's theorem, which is the key for establishing the other sparse versions, may be proved by combining a sparse version of the product argument with rather tricky amalgamation properties of parametersets, as introduced by Frankl, Graham and Rödl [FGR 87].

Not much is known with respect to (m, p, c)-hypergraphs. For $V \subseteq \mathbb{N}$ the edges of the (m, p, c)-*hypergraph* are the (m, p, c)-sets contained in V. Again, using a probabilistic argument Rucinski and Voigt [personal communication] showed that highly chromatic and sparse subgraphs of the (m, p, c)-subgraph on $\{0, \ldots, n-1\}$ exist, but the following conjecture seems to be open.

Conjecture Let g, r, m, p, c be positive integers. Then there exists a set $V \subset \mathbb{N}$ such that the (m, p, c)-hypergraph on V has girth $> g$ and chromatic number $> r$.

In order to conclude this section let us mention that in graph Ramsey theory not only colorings of vertices but also of other subgraphs were investigated [De 73,De 75, EHP 75, NR 79]. The main aspects were on questions of the following type with regard to colorings of edges:

1. Let G be a graph. There exists a graph R such that for every coloring of the edges of R with 2 colors there exists a monochromatic induced copy of G. (Induced version).
2. Let G be a K_m-free graph. There is a K_m-free graph R such that for every coloring of the edges of R with 2 colors there exists a monochromatic subgraph isomorphic to G. (Restricted version)
3. Let G be a graph without cycles of length at most c. For every $c' \geq c$ there exists a graph R without cycles of length at most c' such that for every coloring of the edges of G with two colors there is a monochromatic copy of G. (Sparse version)

In a series of ingeneous but technical papers by [Sp 75,DRV 82,FGR 87,Ne 83,Rö 76,Pr 82,PR 84,PV 85] the question was investigated how and to what extent these theorems in graph theory may be generalized to hypergraphs with arithmetic structure. It turned out that for the above mentioned hypergraphs, a partition theorem holds which simultanously covers the induced, restricted and sparse case. The crucial step in the proof is to show that a product argument covering all these cases holds and then to generalize appropriately the amalgamation techniques which already were successful in Nešetřil and Rödl's papers on this subject.

V. CANONIZING RAMSEY THEORY

Until now we considered equivalence relations with a finite number of classes only. If arbitrarily many classes are admitted at least the injective equivalence relation in which each element is in a class by itself has to be taken into account. So one cannot hope to obtain results (except in trivial cases) stating that at least one class contains something of interest. The problem in canonizing partition theory is shifted towards the description of the invariant patterns which occur for arbitrary equivalence relations. Erdös and Rado [ER 50] initiated the study of the behaviour of structures with respect to arbitrary equivalence relations. The starting point of their investigations was the famous theorem of Ramsey:

Theorem [Ram 30] *Let k, l be positive integers. Then for every equivalence relation with k classes defined on the set $[\mathbb{N}]^l$ of l-element subsets of $\mathbb{N}$ there exists an infinite subset $M \subseteq \mathbb{N}$ such that $[M]^l$ is contained in one class.*

Erdös and Rado allowed arbitrary equivalence relations $\sim$ on $[\mathbb{N}]^l$ and they asked whether subsets $M \subseteq \mathbb{N}$ exist such that $\sim$ behaves in a canonical way on $[M]^l$. It turns out that this can be achieved:

Theorem [ER 50] *Let l be a positive integer. For every equivalence relation on the set $[\mathbb{N}]^l$ of l-element subsets of $\mathbb{N}$ there exists an infinite subset $M \subseteq \mathbb{N}$ and a subset $I \subseteq \{0, 1, \ldots, l-1\}$ with the following property:*

$\{x_0, x_1, \ldots, x_{l-1}\}$ *is equivalent to* $\{y_0, y_1, \ldots, y_{l-1}\}$

iff

$x_i = y_i$ *for all* $i \in I$

for all $\{x_0, x_1, \ldots, x_{l-1}\}, \{y_0, y_1, \ldots, y_{l-1}\} \in [M]^l$ *with* $x_0 < x_1 < \cdots < x_{l-1}$ *and* $y_0 < y_1 < \cdots < y_{l-1}$.

Thus there are 2^l canonical cases describing the invariant equivalence relations and none of these can be omitted without violating the theorem. This may be seen by considering for each $I \subset \{0, \ldots, l-1\}$ the equivalence relation $\sim_I$ which has two sets X, Y in the same class iff X and Y ordered increasingly concide at the elements with index in I. Notice that for $I = \emptyset$ one class contains $[M]^l$, whereas for $I = \{0, 1, \ldots, l-1\}$ the equivalence relation $\sim_I$ is injective, i.e. each element of $[M]^l$ is in a class for itself. Observe that the theorem of Erdös and Rado implies Ramsey's theorem, since for equivalence relations on $[\mathbb{N}]^l$ with a finite number of classes I has to be $\emptyset$.

As usual a *clique* in a graph G is a set of vertices in G any two of them joined by an edge. To a graph G we associate its (l, m)-hypergraph $\bar{G}$: The vertices of $\bar{G}$ are the l-element cliques. A set E of vertices is an edge in $\bar{G}$ iff E is the set of all l element cliques contained in some m element clique of G. So the (l, m) hypergraph $\bar{G}$ of G shows how the m element cliques of G intersect. The following is a finite sparse version of the Erdös-Rado canonization theorem:

Theorem [PV 87] *Let g, l, m be positive integers. Then there exists a graph G such that*
(i) The (l, m)-hypergraph $\bar{G}$ has girth $> g$,
(ii) For every equivalence relation defined on the vertices of $\bar{G}$ there exists an edge E of $\bar{G}$ and $I \subset \{0, \ldots, l-1\}$ such that any two vertices $\{x_0, \ldots, x_{l-1}\}_<$ and $\{y_0, \ldots, y_{l-1}\}_<$ in E are equivalent iff $x_i = y_i$ for $i \in I$.

Next we report on canonical patterns of parameter sets.

Definition Call an equivalence relation $\sim$ on an m-parameter set $P \subseteq A^n$ *invariant* iff there exists an equivalence relation $\approx$ on A such that for every $x, y \in P$ the equivalence $x \sim y$ holds iff x and y coordinatewise are $\approx$ equivalent.

Theorem [PV 83] Let *A be a finite alphabet and $m \in \mathbb{N}$. There exists a least $n = n(A, m) \in \mathbb{N}$ such that for every equivalence relation $\sim$ on A^n there exists an m-parameter subset $P \subset A^n$ on which $\sim$ is invariant.*

Bell numbers count invariant equivalence relations for parameter sets. As a corollary we obtain the canonical version of van der Waerden's theorem on arithmetic progressions. This was proved by Erdös and Graham [EG 80] using Szemeredi's density theorem [Sz 75].

Theorem [EG 80] *Let $\sim$ be an arbitrary equivalence relation defined on $\mathbb{N}$. Then for every k there exists an arithmetic progression P with k terms satisfying (i) or (ii).*
(i) P is contained in one class of $\sim$
(ii) $\sim$ is injective on P.

Proof Let $m = (k-1)^2 + k - 1$, $A = \{0, \ldots, m\}$ and n be large enough. Let $\sim$ be an equivalence relation on $\{0, \ldots, nm\}$. Transfer this to A^n by letting $x \sim^* y$ iff $\sum x_i \sim \sum y_i$ for all $x, y \in A^n$. By the canonical version for parameter sets (n is large enough) there exists an m-parameter subset $P \subset A^n$ on which $\sim^*$ is invariant. Let $I_0, \ldots, I_{m-1} \subset \{0 \ldots n-1\}$ be the sets of active coordinates and $a_i, i \in \{0,\ldots,n-1\}\setminus\bigcup_{j<m} I_j$ be the non-active coordinates of P. Fix $x = \sum a_i$ and $x_j = |I_j|, j = 0,\ldots,m-1$. Transfer the parameter set P back into $\{0,\ldots,nm\}$.

Let $a_j = x + \sum_{l=0}^{m}(j+l)x_l$ for $j = 0,\ldots,k-1$. If the equivalence $\sim$ restricted to the k-term progression $\{a_0 \ldots a_{k-1}\}$ is not injective, then $a_j \sim a'_j$ for some $j < j'$. As $\sim^*$ is invariant on $P \subset A^n$ the associated relation $\approx$ on A which defines the invariant pattern satisfies $j \approx j + l(j'-j)$ for $l = 0,\ldots,k-1$. Thus $\sim$ is constant on the k term progression $x + l(j'-j)x_0, \quad l = 0,\ldots,k-1$.

The analog of this theorem for higher dimensions is quite illustrative too. It is the canonical version of the Gallai–Witt theorem:

Theorem [DGPV 83, PR 86] *Let $S \subset \mathbb{Z}^t$ be finite, and $\sim$ an arbitrary equivalence relation on $\mathbb{Z}^t$. then there exists a linear subspace $U \subseteq \mathbb{Q}^t$ and a homothetic copy $\bar{S} = \bar{a} + \lambda S, (\bar{a} \in \mathbb{Z}^t, \lambda \in \mathbb{N})$ such that the restriction of $\sim$ to $\bar{S}$ is the coset equivalence modulo U, i.e. $\bar{x} \sim \bar{y}$ iff $\bar{x} - \bar{y} \in U$.*

This theorem clearly exhibits the unavoidable invariant patterns of equivalence relations on homothetic copies of S. Note that for $U = \{0\}$ one obtains the injective case and for $U = \mathbb{Q}^t$ the copy $\bar{S}$ is contained in one class of $\sim$. The last case always occurs for equivalence relations with finitely many classes: the so called Gallai–Witt theorem see [Ra 43, Wi 51]. For $t = 1$ one has the canonical van der Waerden theorem.

For (m,p,c)-sets Lefmann investigated arbitrary equivalence relations.

Theorem [Lef 86] *Let $m,p,c \in \mathbb{N}$ and $\sim$ be an equivalence relation on $\mathbb{N}$. Then there exists an (m,p,c)-set M such that the restriction of $\sim$ to M has one of the following patterns:*
(i) $x \sim y$ for all $x,y \in M$
(ii) $x \not\sim y$ for all $x,y \in M, x \neq y$
(iii) $x \sim y$ iff x and y occur in the same row of the defining list (*).

As a corollary one obtains a result which might be called the Rado canonical theorem for partition regular matrices.

Theorem [Lef 86] *Let A be a finite matrix with integral coefficients. Let A have the columns property with blocks $B_0,\ldots,B_{k-1}$. Then for every equivalence relation $\sim$ defined on $\mathbb{N}$ there exists a solution $(x_1,\ldots,x_n)$ of $Ax^T = 0$ such that the restriction of $\sim$ to the solution has one of the following patterns:*
(i) $x_i \sim x_j$ for all $x_i, x_j \in X$
(ii) $x_i \not\sim x_j$ for all $x_i, x_j \in X, x_i \neq x_j$
(iii) $x_i \sim x_j$ iff $i,j \in B_l$ for some l.

This theorem shows that the invariant patterns need not be uniquely determined by A. There obviously are matrices A which have the columns property with various partitions of the columns.

For arbitrary systems of equations the number of unavoidable invariant patterns is determined by the Bell numbers $B_{1+rank(A)}$ [Lef 85]. Regarding Hindman's theorem Taylor proved:

Theorem [Ta 76] Let $\sim$ be an equivalence relation on $\mathbb{N}$. Then there exists an infinite collection x_i of mutually distinct integers such that one of the following cases is valid for all finite nonempty subsets $I, J \subset \mathbb{N}$.
(i) $\sum_I X_i \sim \sum_J X_j$
(ii) $\sum_I X_i \sim \sum_J X_j$ iff $I = J$
(iii) $\sum_I X_i \sim \sum_J X_j$ iff $min I = min J$
(iv) $\sum_I X_i \sim \sum_J X_j$ iff $max I = max J$
(v) $\sum_I X_i \sim \sum_J X_j$ iff $max I = max J$ and $min I = min J$.

In the canonical version of Rado's finite sum theorem cases (iv) and (v) of the above theorem are avoidable. [PV 83].

The canonical situation for systems of equations over Abelian groups was studied in [DL 88].

Theorem [DL 88] *Let G be an Abelian group and A a finite matrix with integral coefficients, which is partition regular over G. Let $B_0, \ldots, B_{k-1}$ be the blocks of the columns property responsible for A being partition regular. Then for every equivalence relation $\sim$ defined on $G \backslash \{0\}$ there exists a solution $(x_1, \ldots, x_n)$ of $Ax^T = 0$ such that the restriction of $\sim$ to the solution has one of the following patterns*
(i) $x_i \sim x_j$ for all $x_i, x_j \in X$
(ii) $x_i \not\sim x_j$ for all $x_i, x_j \in X, x_i \neq x_j$
(iii) $x_i \sim x_j$ iff $i, j \in B_l$ for some l

With this theorem one has a canonizing version for linear homogeneous systems of equations over an Abelian group and thus another highlight in the body of mathematics initiated by Richard Rado in his "Studien zur Kombinatorik".

References

[Ar 70] Arnautov, V.I. (1970). Nondiscrete topologizability of countable rings. Soviet Math. Dokl., 11, 423–426.

[Ba 74] Baumgartner, J. (1974). A short proof of Hindman's theorem, J. Combin. Theory Ser. A. 17, 384–386.

[Br 28] Brauer, A. (1928). Über Sequenzen von Potenzresten. Sitzungsber. Preuß. Akad. Wiss. Math. Phy. Kl., 9–16.

[CEHR 76] Cates, M., Erdös P., Hindman N. and Rothschild B. (1976). Partition theorems for subsets of vector spaces. J. Combin. Theory Ser. A. 20, 279–291.

[CH 75] Cates, M.L. and Hindman, N. (1975). Partition theorems for subspaces of vector spaces. J. Combin. Theory Ser. A. 19, 13–25.

[De 73] Deuber, W. (1973). Generalizations of Ramsey's theorem. in: Infinite and Finite Sets, Colloquia Mathematica Societatis János Bolyai, Keszthely (Hungary), 323–332

[De 73] Deuber, W. (1973). Partitionen und lineare Gleichungssysteme. Math.Z. 133, 109–123.

[De 75] Deuber, W. (1975). Partition theorems for Abelian Groups. J. Combin. Theory Ser. A. 19, 95–108.

[De 75] Deuber, W. (1975 b). Partitionstheoreme für Graphen, Comment.Math.Helvetici 50, 311–320.

[DGPV 83] Deuber, W., Graham, R.L., Prömel H.J. and Voigt, B. (1983). A canonical partition theorem for equivalence relations on Z^t, J. Combin. Theory Ser. A. 34, 331–339.

[DH 87] Deuber, W., Hindman, N. (1987). Partitions and sums of (m,p,c)-sets. Journal of Combinatorial Theory, Ser. A. 45, pp. 300-302.

[DL 88] Deuber, W., Lefmann, H. (1988). Partition regular systems of homogeneous linear equations over Abelian groups: The canonical case. New York Academy of Science.

[DPRV] Deuber, W., Prömel, H.J., Rothschild, B.L. and Voigt, B. (1981). A restricted version of Hales–Jewett's Theorem. in: Finite and Infinite Sets, Colloquia Mathematica Societatis János Bolyai, Eger (Hungary), 231–246.

[DRV 82] Deuber, W., Rothschild, B.L. and Voigt, B. (1982). Induced Partition Theorems. J. Combin. Theory Ser. A. 32, 225–240.

[Er 59] Erdös, P. (1959). Graph theory and probability. Canad. J. Math., 11, 34–38.

[EGMRSS 73] Erdös, P, Graham, R.L., Montgomery, P., Rothschild, B.L., Spencer, J. and Strauss, E. (1973). Euclidean Ramsey Theorems, I, J. Combin. Theory Ser. A 14, 341–363.

[EGMRSS 75] Erdös, P., Graham, R.L., Montgomery, P., Rothschild, B.L., Spencer, J. and Strauss, E. (1975). Euclidean Ramsey Theorems, II. in: Infinite and Finite Sets, ed. A. Hajnal, R. Rado and V.T. Sos. North Holland, Amsterdam, pp. 529–558.

[EGMRSS 75] Erdös, P, Graham, R.L., Montgomery, P., Rothschild, B.L., Spencer, J. and Strauss, E. (1975). Euclidean Ramsey Theorems, III. in: Infinite and Finite Sets, ed. A. Hajnal, R. Rado and V.T. Sos. North Holland, Amsterdam, pp. 559–584.

[EG 75] Erdös, P. and Graham, R.L. (1975). On partition theorems for finite graphs. in: Finite and Infinite Sets. (A. Hajnal, R. Rado and V.T. Sos eds.) Colloq. Math. Soc. Janos Bolyai 10, North Holland, Amsterdam, 515–527.

[EG 80] Erdös, P. and Graham, G. (1980). Old and New Problems and Results in Combinatorial Number Theory. L'Enseigment Mathematique Monographie No. 28, Genéve.

[EH 66] Erdös, P. and Hajnal, A. (1966). On chromatic number of graphs and set systems. Acta. Math. Acad. Sci. Hungar.17, 61–99.

[EHP 75] Erdös, P., Hajnal, A. and Pósa, L. (1975). Strong embeddings of graphs into colored graphs, in: Infinite and Finite Sets, Vol. 1 (A. Hajnal, Ed.), Elsevier, Amsterdam, 585–596.

[ER 50] Erdös, P. and Rado, R. (1950). A Combinatorial Theorem. J. London Math. Soc. 25, 249–255.

[ES 35] Erdös, P. and Szekeres, G. (1935). A combinatorial problem in geometry. Composito Math. 2, 463–470.

[FR 87] Frankl, P. and Rödl, V. (1987). A Partition Property of Simplices. preprint.

[FGR 87] Frankl, P, Graham, R.L. and Rödl, V. (1987). Induced restricted Ramsey theorems for spaces. J. Combin. Theory Ser. A.44, 120–128.

[FW 78] Fürstenberg, H. and Weiss, B. (1978). Topological Dynamics and Combinatorial Number Theory. J. Anal. Math. 34, 61–85.

[GLR 72] Graham, R.L., Leeb, K. and Rothschild, B.L. (1972). Ramsey's theorem for a class of categories. Adv. in Math. 8, 417–433.

[GR 71] Graham, R.L. and Rothschild, B.L. (1971). Ramsey's theorem for n-parameter sets. Trans. AMS 159, 257–292.

[HJ 63] Hales, A.W. and Jewett, R.I. (1963). Regularity and positional games. Trans. Amer. Math. Soc. 106, 222–229.

[Hi 74] Hindman, N. (1974). Finite sums from sequences within cells of a partition of N. J. Combin. Theory Ser. A. 17, 1–11.

[Le 73] Leeb, K. (1973). Vorlesungen über Pascaltheorie. Arbeitsberichte des Instituts für Math. Maschinen und Datenverarbeitung, Friedrich Alexander Universität Erlangen–Nürnberg, 6.

[Le 75] Leeb, K. (1975). A full Ramsey theorem for the Deuber category. in: Infinite and Finite Sets, (A. Hajnal, R. Rado, V.T. Sós eds.), Colloq. Math. Soc. János Bolyai 10, North Holland, Amsterdam, 1043–1049.

[Lef 85] Lefmann, H. (1985). Kanonische Partitionssätze. Dissertation. Bielefeld

[Lef 86] Lefmann, H. (1986). A canonical version for partition regular systems of linear equations. J. Combin. Theory Ser. A. 41, 95–104.

[Lef 88] Lefmann, H. (1988). On partition regular systems of equations. preprint.

[Lo 68] Lovász, L. (1968). On chromatic number of finite set-systems, Acta. Math. Acad. Hungar. 19, 59–67.

[My 55] Mycielski, J. (1955). Sur le coloriage des graphes. Colloq. Math. 3, 161–162.

[Ne 66] Nešetřil, J. (1966). k-chromatic graphs without cycles of length at most 7. (in Czech), Comment. Math. Univ. Carolinae 7, 373–376.

[NR 79] Nešetřil, J. and Rödl, V. (1979). A short proof of highly chromatic graphs without short cycles. J. Combin. Theory Ser. B.27, 225–227.

[NR 79] Nešetřil, J. and Rödl, V. (1979). Partition theory and its applications. in: Surveys in Combinatorics, Cambridge Univ. Press, Cambridge, 96–156.

[NR 79] Nešetřil, J. and Rödl, V. (1979). A short proof of highly chromatic graphs without short cycles. J. Combin. Theory Ser. B. 27, 225–227.

[NR 83] Nešetřil, J. and Rödl, V. (1983). Ramsey classes of set systems. J. Combin. Theory Ser. A. 34, 183–201.

[NR 86] Nešetřil, J. and Rödl, V. (1986). On sets of integers with the Schur property. Graphs and Combinatorics 2, 269–275.

[PH 77] Paris, J. and Harrington, L. (1977). A mathematical incompleteness in Peano arithmetic, in: Handbook of Mathematical Logic (J. Barwise ed.) North Holland Amsterdam, 1133–1142.

[Pr 82] Prömel, H.J. (1982). Induzierte Partitionssätze. Dissertation. Bielefeld

[PR 86] Prömel, H.J. and Rödl, V. (1986). An elementary proof of the canonizing version of Gallai–Witt's theorem. J. Combin. theory Ser. A. 42, 144–149.

[PR 87] Prömel, H.J. and Rothschild, B.L. (1987). A canonical restricted version of van der Waerden's Theorem. Combinatorica 7, 115–119.

[PV 83] Prömel, H.J. and Voigt, B. (1983). Canonical partition theorems for parameter sets. J. Combin. theory Ser. A. 35, 309–327.

[PV 85] Prömel, H.J. and Voigt, B. (1985). Canonizing Ramsey Theorems for finite graphs and hypergraphs. Discrete Math. 54, 49–59.

[PV 85a] Prömel, H.J. and Voigt, B. (1985). Graham Rothschild parameter sets. Report No. WP 85403, Institut für O.R., Universität Bonn, to appear in: The mathematics of Ramsey Theory (eds) Nešetřil, V. Rödl, Springer.

[PV 87] Prömel, H.J. and Voigt, B. (1987). Ramsey Theory for finite graphs II. Report No. 87457–OR, Institut für Diskrete

Mathematik, Universität Bonn.

[PV 88] Prömel, H.J. and Voigt, B. (1988). A Sparse Graham–Rothschild Theorem. Transactions of the American Mathematical Society, vol. 309, no. 1. 113–137.

[Ra 33] Rado, R. (1933). Verallgemeinerung eines Satzes von van der Waerden mit Anwendungen auf ein Problem der Zahlentheorie. Sitzungsber. Preuß. Akad. Wiss. Phys.- Math. Klasse 17, 1–10.

[Ra 33] Rado, R. (1933). Studien zur Kombinatorik. Math. Z. 36, 424–480.

[Ra 43] Rado, R. (1943). Note on combinatorial analysis. Proc. London Math. Soc. 48, 122–160.

[Ra 69] Rado, R. (1969). Some partition theorems. in: Combinatorial Theory and its Applications (P. Erdös, A. Rényi and V.T. Sós eds.), Colloq. Math. Soc. János Bolyai 4, North Holland, Amsterdam, 929–936.

[Ram 30] Ramsey, F.P. (1930). On a Problem of Formal Logic. Proc. London Math. Soc. 30, 264–286.

[Rö 76] Rödl, V. (1976). A generalization of Ramsey's theorem. in: Graphs, Hypergraphs and Block Systems, Yielona Gora, 211–220.

[Rö 81] Rödl, V. (1981). On Ramsey families of sets. unpublished manuscript.

[Sa 69] Sanders, J. (1969). A generalization of Schur's theorem. Dissertation, Yale University.

[Sch 16] Schur, I. (1916). Über die Kongruenz $x^m + y^m \equiv z^m (mod p)$. J.ber. d. Dt. Math.-Verein, 114–117.

[Sh 88] Shelah, S. (1988). Primitive recursive bounds for van der Waerden numbers, Journal of the American Mathematical Society, vol. 1, 683–697.

[Sp 75] Spencer, J. (1975). Restricted Ramsey configurations. J. Combin. Theory Ser. A. 19, 278–286.

[Ta 76] Taylor, A.D. (1976). Canonical partition relation for finite subsets of ω. J. Combin. Theory Ser. A. 21, 137–146.

[vdW 27] van der Waerden, B.L. (1927). Beweis einer Baudetschen Vermutung. Nieuw Arch. Wisk. 15, 212-216.

[Wi 51] Witt, E. (1951). Ein Kombinatorischer Satz der Elementargeometrie. Math. Nachr. 6, 261-262.

[Zy 49] Zykov, A.A. (1949). On some properties of linear complexes. (in Russian), Mat. Sbernik N.S. 24,, 163–188.; Amer. Math. Soc.Transl. 79. 1952.

Walter A. Deuber, Fakultät für Mathematik, Universität Bielefeld, 4800 Bielefeld

DESIGNS AND AUTOMORPHISM GROUPS

Jean Doyen

INTRODUCTION

In this paper D will always denote a t-(v,k,λ) design (or, in the "Belgian" notation, an $S_\lambda(t,k,v)$) where we assume $t < k < v$ to avoid trivial examples. Aut D will as usual denote the full automorphism group of D.

Like many other mathematical structures satisfying weak axioms (for example graphs), designs are much too wild to be classified. Moreover, it is part of folklore to say that almost all designs are "ugly" (i.e. have no automorphism other than the identity). This was proved rigorously in the case of 2-(v,3,1) designs by Babai (1980).

Therefore it is quite natural to restrict our attention to "nice" designs (i.e. those designs whose automorphism group satisfies some reasonable transitivity hypothesis) and try to classify them. Before going any further, here is a typical example of such a classification:

Theorem (Kantor 1985). If Aut D is t-transitive on the points of a t-(v,k,1) design with $t \geq 3$ then D is one of the following:
(1) the $3\text{-}(2^d,4,1)$ design whose points and blocks are the points and planes of the affine space AG(d,2) with $d \geq 3$,
(2) the $3\text{-}(q^e+1, q+1, 1)$ design whose points and blocks are respectively the elements of $GF(q^e) \cup \{\infty\}$ and all the images of $GF(q) \cup \{\infty\}$ under $PGL(2,q^e)$, with q a prime power and $e \geq 2$, or
(3) one of the Mathieu designs 4-(11,5,1), 5-(12,6,1), 3-(22,6,1), 4-(23,7,1), 5-(24,8,1).

More generally one would like to classify all pairs (G,D) consisting of a t-design D and a subgroup G of Aut D where G satisfies some given transitivity assumption as for example transitivity on points, transitivity on blocks, primitivity on points, primitivity on blocks, transitivity on flags (i.e. on incident point-block pairs), n-transitivity on points, n-homogeneity on points, and so on.

Of course, these assumptions are not independent of each other. The following results exhibit some of the relations between them:

Theorem (Block 1967). If $t \geq 2$ and if G is block-transitive, then G is point-transitive.

Note that this result has an immediate interesting consequence:

Corollary. If $t \geq 3$ and if G is flag-transitive, then G is 2-transitive on points.

Proof: For any point x of the t-design D, the stabilizer G_x is transitive on the blocks of the derived (t-1)-design D_x and so G_x is transitive on the points of D_x by Block's theorem. It follows that G is 2-transitive on the points of D.

Block's theorem has been generalized later as follows:

Theorem (Kreher 1986). If G is block-transitive on a t-design D, then G is $\lfloor t/2 \rfloor$-homogeneous on the points of D.

For $t \geq 4$ it follows that block-transitivity implies 2-homogeneity on points (hence primitivity on points). For $t < 4$ block transitivity does not necessarily imply point-primitivity. To get an infinite family of counter examples, consider the 2-design D consisting of the points and hyperplanes of a Desarguesian projective space $PG(d,q)$ where $d \geq 2$ and $(q^{d+1}-1)/(q-1)$ is not a prime, and take for G the group generated by a Singer cycle.

The following result shows that the counter examples have necessarily a small number of points (with respect to k):

Theorem (Delandtsheer & Doyen 1989a). In a t-(v,k,λ) design D with $t \geq 2$ and $v > [\, k(k-1)/2 - 1\,]^2$ any block-transitive automorphism group G is point-primitive.

We have already seen that for $t \geq 3$ flag-transitivity implies 2-transitivity on points (hence point-primitivity). Counter examples described by Davies (1987) show that this is no longer true for $t = 2$. However, several sufficient conditions on the parameters of a 2-(v, k, λ) design are known for which the implication holds. As was proved by Kantor (1969) for example, it suffices to have $(r, \lambda) = 1$ (where r denotes as usual the number of blocks through a given point). Later we will use a particular case of this result:

Theorem (Higman & McLaughlin 1961). In any 2-(v,k,1) design flag-transitivity implies point-primitivity.

Finally, we mention a useful and surprising result:

Theorem (Camina & Gagen 1984). In any 2-(v,k,1) design where k divides v, block-transitivity implies flag-transitivity.

FLAG-TRANSITIVE 2-(v,k,1) DESIGNS

Let D be a 2-(v,k,1) design - from now on the blocks of D will be called lines - and let G be a subgroup of Aut D.

The pairs (G,D) where G is 2-transitive on the points of D have been classified by Kantor (1985). The proof uses the classification of finite 2-transitive permutation groups, which itself relies on the classification of finite simple groups.

A weakening of this transitivity hypothesis leads to the problem of classifying all pairs (G,D) where G is 2-homogeneous on the points of D. This problem was solved by Delandtsheer et al. (1986). An interesting class of designs, the so-called Netto systems N(q), arises from this investigation. For each prime power q congruent to 7 mod 12 there is (up to isomorphism) exactly one Netto system N(q) on q points which can be defined as follows. Let ε be a primitive sixth root of unity in GF(q) and let $A\Gamma^2L(1,q)$ denote the group of all permutations of GF(q) of the form $x \to a^2x^\sigma + b$ with a, b in GF(q), $a \neq 0$ and σ in Aut(GF(q)). The points of N(q) are the elements of GF(q) and the lines are the image of $\{0,1,\varepsilon\}$ under $A\Gamma^2L(1,q)$. Clearly N(7) is isomorphic to PG(2,2) whose full automorphism group is 2-transitive on points. If $q > 7$ Robinson (1975) proved that Aut N(q) is isomorphic to $A\Gamma^2L(1,q)$ which is 2-homogeneous but not 2-transitive on points. Now comes a remarkable fact: the Netto systems N(q) with $q > 7$ are the only 2-(v,k,1) designs whose full automorphism group is 2-homogeneous but not 2-transitive on points.

A further weakening of 2-homogeneity leads to the problem of classifying all pairs (G,D) where G is flag-transitive on D. This is not yet completely solved. However, some progress has been made recently on this classification problem and we will now make a quick survey of some of the basic ideas involved and of the main results already obtained.

So let a group $G \leq \mathrm{Aut}\, D$ act flag-transitively on a 2-(v,k,1) design D. From the assumption $2 < k < v$ it follows easily that $r > \sqrt{v}$.

Lemma 1. For any point x of D we have (i) r divides $(|G_x|, v-1)$ and (ii) $|G_x| > \sqrt[3]{|G|}$ (cube root bound).

Proof: Property (i) follows from the fact that G_x acts transitively on the r lines through x and from the equality $v-1 = r(k-1)$. On the other hand, $\sqrt{v} < r$ forces $v < (|G_x|, v-1)^2 \leq |G_x|^2$ and since $|G| = v\,|G_x|$ we get property (ii).

Lemma 2. G_x is a maximal subgroup of G.

Proof: By the theorem of Higman and McLaughlin G acts primitively on the points of D and so the stabilizer of any point x is a maximal subgroup of G (a well-known property of primitive permutation groups).

Now, given an abstract group G, the strategy for constructing all 2-(v,k,1) designs D on which G acts as a flag-transitive automorphism group is clear. For each maximal subgroup of G satisfying the cube root bound, let v be its index in G and compute the possible values of r by Lemma 1(i), keeping in mind that $r > \sqrt{v}$. The value of k is given by $v-1 = r(k-1)$.

Of course, there is still an important question to answer. Which groups G should we start with?

Here again the fact that G is point-primitive is very helpful. A fundamental theorem due to O'Nan and Scott reduces the finite primitive permutation groups to five major types (for more details, see for example Liebeck et al. (1988)). It turns out that G is restricted to only two of these five types:

Theorem (Buekenhout et al. 1988). If $G \leq \mathrm{Aut}\, D$ is flag-transitive on a 2-(v,k,1) design D, then one of the following holds:
(1) Affine case: G has an elementary abelian minimal normal subgroup T, acting regularly on the points of D (and so $v = p^d$ for some prime p), or
(2) Almost simple case: G has a non-abelian normal simple subgroup N such that $N \leq G \leq \mathrm{Aut}\, N$.

Note that in the affine case $G = T\,G_0$ where T is the group of translations of some affine space AG(d,p) (p prime) left invariant by G so that $G_0 \leq GL(d,p)$.

Using the classification of finite simple groups a team of six people (F. Buekenhout, A. Delandtsheer, J. Doyen, P. Kleidman, M. Liebeck and J. Saxl) decided to attack these two cases. The affine case seems to be the most difficult to handle completely and is still under study. In the almost simple case a complete list of flag-transitive pairs (G,D) has been obtained recently. Before stating this result we need to introduce a few classes of examples:

A unital of order n is a $2\text{-}(n^3+1,\ n+1,\ 1)$ design. For any prime power q, $U_H(q)$ denotes the Hermitian unital of order q whose points and lines are respectively the absolute points and non-absolute lines of a unitary polarity in $PG(2,q^2)$, the incidence being the natural one. The full automorphism group of $U_H(q)$ is $P\Gamma U(3,q)$.

For any $q = 3^{2e+1}$, $(e \geq 0)$, $U_R(q)$ denotes the Ree unital of order q. Its points and lines are respectively the Sylow 3-subgroups and the involutions of the Ree group ${}^2G_2(q)$, a point and a line being incident if and only if the involution normalizes the Sylow 3-subgroup (Lüneburg 1966). The full automorphism group of $U_R(q)$ is Aut ${}^2G_2(q)$.

For any $q = 2^e$ $(e \geq 3)$, W(q) denotes the $2\text{-}(\ q(q-1)/2,\ q/2,\ 1\ )$ design whose points are the lines of PG(2,q) disjoint from a given complete conic C (i.e. the union of an irreducible conic and its nucleus) and whose lines are the points of PG(2,q) outside C, the incidence being the natural one. The full automorphism group of W(q) is $P\Gamma L(2,q)$.

Main Theorem (Buekenhout, Delandtsheer, Doyen, Kleidman, Liebeck and Saxl, to appear). If $G \leq$ Aut D is flag-transitive on a $2\text{-}(v,k,1)$ design D, then one of the following holds:

(1) G is of affine type, or

(2) $N \leq G \leq$ Aut N for some non-abelian normal simple group N, and the only possibilities are :

Case P: D = PG(d,q), $d \geq 2$ and N = PSL(d+1,q) or D = PG(3,2) and N= Alt(7),
Case H: D = $U_H(q)$ and N = PSU(3,q),
Case R: D = $U_R(q)$, $q = 3^{2e+1}$, $(e \geq 1)$ and $N = {}^2G_2(q)$,
Case W: D = W(q), $q = 2^e$, $(e \geq 3)$ and N = PSL(2,q).

(Note that the smallest Ree unital $U_R(3)$ is isomorphic to $W(8)$ and so appears in case W).

In order to illustrate the difficulties involved in an analogous complete treatment of the affine case we remark that those flag-transitive projective planes which are unknown at present should appear in the affine case. This is a consequence of the following strong result:

Theorem (Kantor 1987). If G is a flag-transitive automorphism group of a finite projective plane P of order n, then
(1) $P = PG(2,q)$ and $G \geq PSL(3,q)$, or
(2) n^2+n+1 is a prime and G is a sharply flag-transitive Frobenius group of order $(n^2+n+1)(n+1)$.

Note also that the Netto systems $N(q)$ appear in the affine case.

TWO APPLICATIONS

The purpose of this section is to convince the reader that, although a complete list of flag-transitive pairs (G,D) is not yet available in the affine case, the main theorem stated above can already be applied successfully to a series of problems.

Problem 1: Line-transitive unitals. Let D be a unital of order n, that is a $2\text{-}(n^3+1, n+1, 1)$ design. What can be said about D if there is a group $G \leq \mathrm{Aut}\, D$ acting line-transitively on D?

Since $k = n+1$ divides $v = n^3+1$, G is flag-transitive on D by Camina and Gagen (1984), and so G is of affine type or is almost simple by the main theorem.

If G is of affine type, then $v = p^d$ for some prime p. But the Diophantine equation $n^3+1 = p^d$ has only one solution, namely $n = 2$, $p = 3$ and $d = 2$. Therefore D is the Hermitian unital $U_H(2)$ isomorphic to $AG(2,3)$.

If G is almost simple, it is an easy exercise to prove that D cannot be isomorphic to the design of points and lines of $PG(d,q)$ and that D is isomorphic to $W(q)$ only if $q = 8$. Therefore by the main theorem D is $U_H(q)$ or $U_R(q)$ (remember that $W(8)$ is isomorphic to $U_R(3)$). Thus we have proved the following

Theorem. The only line-transitive unitals are the Hermitian unitals and the Ree unitals.

In particular, any unital whose order is not a prime power (like the unital of order 6 constructed by Mathon (1987)) cannot have a line-transitive automorphism group.

Problem 2: Line-transitive maximal (v,k)-arcs. A maximal (v,k)-arc of a d-dimensional finite projective space P of order n is a nonempty set S of v points of P such that every line of P meets S in 0 or in k points (where we assume $2 \le k \le n$ to avoid trivial examples).

Theorem (Tallini Scafati 1967). If $d \ge 3$ the only maximal (v,k)-arcs of PG(d,q) are the affine subspaces AG(d,q).

The situation is much more interesting when $d = 2$. Let S be a maximal (v,k)-arc of a (not necessarily Desarguesian) finite projective plane P of order n. The lines of P intersecting S induce a 2-(v,k,1) design on S, where $v = (n+1)(k-1) + 1$. The lines of P containing a point x outside S and intersecting S determine a partition of the v points of this design into lines of size k, and so $k \mid v = nk+k-n$; it follows that $k \mid n$. This necessary condition is well-known not to be sufficient for the existence of a maximal (v,k)-arc: for example, there is no maximal (v,3)-arc in $PG(2,3^e)$ if $e \ge 2$ as was proved by Thas (1975). However, Denniston (1969) showed that the condition $k \mid 2^e$ is sufficient in $PG(2,2^e)$.

A maximal (v,k)-arc S in P is line-transitive if there is a group $G \le \mathrm{Aut}\, P$ stabilizing S and acting transitively on the lines of S. Clearly, any line-transitive affine plane A of order n is a line-transitive maximal (n^2,n)-arc in the projective plane P obtained by adjoining to A its points at infinity (A is necessarily a translation plane, and so n is a prime power). On the other hand, if C is the union of an irreducible conic and its nucleus in $PG(2,2^e)$ with $e \ge 3$, the set of all lines disjoint from C is a maximal $(2^{e-1}(2^e - 1), 2^{e-1})$-arc S in the dual projective plane and the stabilizer $PSL(2,2^e)$ of C in $PGL(3,2^e)$ acts transitively on the lines of S. We will use the same notation as above and denote this maximal arc by $W(2^e)$.

Since $k \mid v$, any line-transitive maximal (v,k)-arc is also flag-transitive by Camina and Gagen (1984). The following result is then a (non trivial) consequence of the main theorem:

Theorem (Delandtsheer & Doyen 1989b). Let P be a (not necessarily Desarguesian) finite projective plane of order n. If $G \leq \mathrm{Aut}\, P$ stabilizes a maximal (v,k)-arc S of P and acts transitively on the lines of S, then
(1) $n = p^e$ (p prime, $e \geq 1$), $k = n$, S is a line-transitive translation affine plane A and G contains the translation group of A, or
(2) $n = 2^e$, ($e \geq 3$), $k = n/2$, $S = W(2^e)$ in $PG(2,2^e)$ and $PSL(2,2^e) \leq G \leq P\Gamma L(2,2^e)$, or
(3) $n = 4$, $k = 2$, S is a hyperoval in PG(2,4) and G = Alt(6) or Sym(6).

The case $k = 2$ of this theorem was already proved by Abatangelo (1986).

REFERENCES

Abatangelo, V. (1986). Doubly transitive (n+2)-arcs in projective planes of even order. J. Combin. Theory Ser. A 42, 1 - 8.

Babai, L. (1980). Almost all Steiner triple systems are asymmetric. Ann. Discrete Math. 7, 37 - 39.

Block, R. E. (1967). On the orbits of collineation groups. Math. Z. 96, 33 - 49.

Buekenhout, F., Delandtsheer, A. & Doyen, J. (1988). Finite linear spaces with flag-transitive groups. J. Combin. Theory Ser. A 49, 268 - 293.

Camina, A. R. & Gagen, T. M. (1984). Block transitive automorphism groups of designs. J. Algebra 86, 549 - 554.

Davies, H. (1987). Flag-transitive and primitivity. Discrete Math. 63, 91 - 93.

Delandtsheer, A., Doyen, J., Siemons, J. & Tamburini, C. (1986). Doubly homogeneous 2-(v,k,1) designs. J. Combin. Theory Ser. A 43, 140 - 145.

Delandtsheer, A. & Doyen, J. (1989a). Most block-transitive t-designs are point-primitive, to appear in Geom. Dedicata.

Delandtsheer, A. & Doyen, J. (1989b). A classification of line-transitive maximal (v,k)-arcs in finite projective planes, to appear.

Denniston, R. H. F. (1969). Some maximal arcs in finite projective planes. J. Combin. Theory Ser. A 6, 317 - 319.

Higman, D. G. & McLaughlin, J. E. (1961). Geometric ABA-groups. Illinois J. Math. 5, 382 - 397.

Kantor, W. M. (1969). Automorphism groups of designs. Math. Z. 109, 246 - 252.

Kantor, W. M. (1985). Homogeneous designs and geometric lattices. J. Combin. Theory Ser. A 38, 66 - 74.

Kantor, W. M. (1987). Primitive permutation groups of odd degree, and an application to finite projective planes. J. Algebra 106, 15 - 45.

Kreher, D. (1986). An incidence algebra for t-designs with automorphisms. J. Combin. Theory Ser. A 42, 239 - 251.

Liebeck, M. L., Praeger, C. E. & Saxl, J. (1988). On the O'Nan-Scott theorem for finite primitive permutation groups. J. Australian Math. Soc. Ser. A 44, 389 - 396.

Lüneburg, H. (1966). Some remarks concerning the Ree groups of type (G_2). J. Algebra 3, 256 - 259.

Mathon, R. (1987). Constructions for cyclic Steiner 2-designs. Ann. Discrete Math. 34, 353 - 362.

Robinson, R. M. (1975). The structure of certain triple systems. Math. Comp. 29, 223 - 241.

Tallini Scafati, M. (1967). Caratterizzazione grafica delle forme hermitiane di un $S_{r,q}$. Rend. Mat. 26, 273 - 303.

Thas, J. A. (1975). Some results concerning {(q+1)(n-1), n}-arcs and {(q+1)(n-1)+1, n}-arcs in finite projective planes of order q. J. Combin. Theory Ser. A 19, 228 - 232.

Département de Mathématiques, Campus Plaine CP 216,
Université Libre de Bruxelles, B-1050 Bruxelles, Belgium

ON MATCHINGS AND HAMILTON CYCLES IN RANDOM GRAPHS

A.M. Frieze

§1. Introduction

The aim of this paper is to review what is currently known about large matchings and cycles in random graphs, in particular perfect matchings and Hamilton cycles. It may seem odd to treat these two subjects together, but we will see in §2 that proofs of theorems on the two topics can be similar.

We will first discuss the two basic results concerning $G_{n,m}$. This graph has vertex-set $[n] = \{1,2,\dots,n\}$ and edge-set $E_{n,m}$ which is a random m-subset of the $N = \binom{n}{2}$ possibilities.

We start with perfect matchings and a theorem of Erdös and Rényi [1966], the "founding fathers" of the subject of random graphs. As usual let $\delta(G)$ denote the minimum degree of graph G.

Theorem 1.1. Let $m = \frac{1}{2} n(\log n + c_n)$. Then

$$\lim_{\substack{n\to\infty \\ n \text{ even}}} \Pr(G_{n,m} \text{ has a perfect matching}) = \lim_{\substack{n\to\infty \\ n \text{ even}}} \Pr(\delta(G_{n,m}) \geq 1)$$

$$= \begin{cases} 0 & c_n \to -\infty \\ e^{-e^{-c}} & c_n \to c \\ 1 & c_n \to +\infty . \end{cases}$$

(When n is odd $G_{n,m}$ has a matching of size $\lfloor n/2 \rfloor$ with the same limiting probability that there is at most one isolated vertex).

Their proof is complicated and based on Tutte's theorem [Tutte,

1947] for the existence of perfect matchings. We give an outline of a relatively simple proof in §2.

This theorem sets the scene nicely. A simple necessary condition ($\delta \geq 1$) is nearly always sufficient. One can guess that $\delta \geq 2$ is nearly always sufficient for the existence of Hamilton cycles. In fact we have the following theorem of Komlós and Szemerédi [1983].

Theorem 1.2. Let $m = \frac{1}{2} n (\log n + \log\log n + c_n)$. Then

$$\lim_{n\to\infty} \Pr(G_{n,m} \text{ is Hamiltonian}) = \lim_{n\to\infty} \Pr(\delta(G_{n,m}) \geq 2)$$

$$= \begin{cases} 0 & c_n \to -\infty \\ e^{-e^{-c}} & c_n \to c \\ 1 & c_n \to +\infty \end{cases}$$

Theorem 1.2 took somewhat longer to prove than Theorem 1.1. There were several interim results showing that if m grows large enough then almost every $G_{n,m}$ is Hamiltonian. Amongst these the most important is probably that of Posa [1976] showing that $m = Kn\log n$ for sufficiently large K suffices. The paper contains a result (see Lemma 2.2) which is the foundation for many proofs of Hamiltonicity. Also Korsunov [1976] claimed a proof for $c_n \to \infty$ in an extended abstract.

Erdös and Rényi [1961] envisaged a *graph process* in which a random graph grows one edge at a time, the additional edge being chosen randomly from the edges remaining. Let $\Gamma_0, \Gamma_1, \ldots, \Gamma_m, \ldots, \Gamma_N$ denote the random sequence of graphs produced. Let $m_k^* = \min\{m: \delta(\Gamma_m) \geq k\}$. Now clearly, $\Gamma_{m_2^*-1}$ is not Hamiltonian and it is rather nice that the following strengthening of Theorem 1.2 is possible:

Theorem 1.3. $\lim_{n\to\infty} \Pr(\Gamma_{m_2^*} \text{ is hamiltonian}) = 1.$

This theorem was claimed in [Komlós and Szemeredi, 1983] without proof as a reformulation of Theorem 1.2. This is not quite true and Bollobás gave a complete proof in [Bollobás, 1984]. Subsequently, Ajtai, Komlós and Szemerédi [1985] gave a different proof. If Theorem 1.3 is true one should not be surprised with

Theorem 1.4. $\lim_{\substack{n\to\infty \\ n \text{ even}}} \Pr(\Gamma_{m_1^*} \text{ has a perfect matching}) = 1.$

For a proof see Bollobás [1985] (Theorem VII.22).

In §2 we will give outline proofs of Theorem 1.1 and 1.2. They demonstrate the general approach to this topic. In §3 we give generalizations of these theorems. In §4-6 we discuss other models of random graphs, in §7 we discuss random digraphs and in §8 we end with some open problems.

§2."Proofs" of Theorems 1.1 and 1.2

We first consider Theorem 1.1. Suppose first that n is even, $m = \frac{1}{2}n(\log n + c)$, and $G_{n,m}$ has no perfect matching. Let $X = \{x\colon x$ is left exposed by some maximum matching$\}$. For $x \in X$ let $Y(x) = \{y\colon$ $\exists$ a maximum matching which leaves x and y exposed$\}$.

For $S \subseteq [n]$ let $N(S) = \{t \notin S\colon \exists\, s \in S$ such that $st \in E_{n,m}\}$.

Lemma 2.1. $|N(Y(x))| < |Y(x)|$ for $x \in X$.

Proof: Suppose $x \in X$. Fix a maximum matching M leaving v exposed. Let S be the remaining set of vertices left exposed by M. Then $y \in Y = Y(x)$ iff there exists an even length alternating path P_y from $s \in S$ to y.

Suppose now that $zy \in E_{n,m}$ with $y \in Y$, $z \notin Y$. The lemma follows from the claim that there exists $y' \in Y$ such that $y'z \in M$. To see this note that either $z \in P_y$ or $P_y + yz + zy'$ is an even length alternating path, where zy' is the edge of M covering z. □

Now clearly

(2.1) $$x \in X, \quad y \in Y(x) \text{ implies } xy \notin E_{n,m}.$$

Now let a graph with vertex set $[n]$ be in EX_k if $S \subseteq [n]$, $|S| \le \frac{n}{k+2}$ implies $|N(S)| \ge k|S|$.

Now it is not difficult to show that a.e. $G_{n,m}$ with $\delta \ge 1$ satisfies

(2.2) $G_{n,m} - A \in EX_1$ for all matchings A which avoid vertices of degree 1.

(The proof of this is through a somewhat tedious calculation where it is useful to consider $[n]$ partitioned into small vertices of degree at most $\frac{1}{10} \log n$ and large vertices - for details see similar calculations in Bollobás, Fenner and Frieze [1987]). We exploit the existence of this "hole" in $G_{n,m}$ implied by (2.1) and (2.2).

Observe also that a.e. $G_{n,m}$ has fewer than $\log n$ vertices of degree 1 and $\Delta \le 3 \log n$. We can now finish the proof fairly quickly using an argument based on that in Fenner and Frieze [1983] (the so called "colouring argument"). Let $\mathcal{G}(n,m)$ denote the set of all m edge graphs on $[n]$ and $\mathcal{G}_1(n,m)$ those which have $\delta \geq 1$, no perfect matching, fewer than $\log n$ vertices of degree 1, $\Delta \leq 3 \log n$ and satisfy (2.2). Let $\omega = \lceil \log n \rceil$ and A be a random ω-subset of $E_{n,m}$. Let $\mathcal{A}$ be the event {(i) A avoids at least one maximum matching of $G_{n,m}$, (ii) A avoids vertices of degree 1 and (iii) A is a matching}. Then

$$\Pr(\mathcal{A} \mid G_{n,m} \in \mathcal{G}_1(n,m)) \geq \frac{1}{3}.$$

(A simple calculation once one has fixed $G \in \mathcal{G}_1(n,m)$). Hence

$$\Pr(G_{n,m} \in \mathcal{G}_1(n,m)) \leq 3 \Pr(\mathcal{A}).$$

But

$$\Pr(\mathcal{A}) = \sum_H \Pr(\mathcal{A} \mid G_{n,m} - A = H)\Pr(G'_{n,m} - A = H)$$

$$\leq \sum_{H \in EX_1} \Pr(A \cap F_H = \phi \mid G_{n,m} - A = H)\Pr(G_{n,m} - A = H)$$

where $F_H = \{xy: x \in X, y \in Y(x)\}$ for some sets $X, Y(x)$, $x \in X$ defined in terms of H only. $A \cap F_H = \phi$ follows from $\mathcal{A}$(i) and $H \in EX_1$ follows from $\mathcal{A}$(ii), $\mathcal{A}$(iii) and (2.2). But if H is fixed then A is a random ω-subset of $\overline{E(H)}$. Hence

$$\Pr(\mathcal{A}) \leq \sum_{H \in EX_1} \left(\frac{9}{10}\right)^{\omega} \Pr(G_{n,m} - A = H) \leq \left(\frac{9}{10}\right)^{\omega}$$

and the theorem follows after tidying up.

Now let us turn to Theorem 1.2. We will see that in some sense we can prove this theorem by replacing 1 by 2 in the relevant places. Suppose $G = G_{n,m}$ is not hamiltonian. Let $X = \{x: x$ is an endpoint of some longest path of $G\}$. For each $x \in X$ choose some longest path $P_x = (x = x_0, x_1, x_2, \ldots, x_p)$ with x as one endpoint. Suppose that the edge $x_p x_i$, $i \leq p-2$, exists. A *rotation* with x as fixed endpoint creates the new longest path $(x_0, x_1, \ldots, x_i, x_p, x_{p-1}, \ldots, x_{i+1})$, also having x as one endpoint. Let now $Y(x)$ be the set of vertices which are the endpoints other than x of those longest paths which can be obtained from P_x by a sequence of rotations with x as fixed endpoint.

In place of Lemma 2.1 we have

Lemma 2.2 (Posa [1976]). $|N(Y(x))| < 2|Y(x)|$ for $x \in X$.

Proof: If $z \in N(Y(x))$ then z is the neighbour, on P_x, of some vertex in $Y(x)$. □

Clearly $|X| > |Y(x)|$ for all $x \in X$ and furthermore (2.1) holds if G is connected (using the current definition for X, Y). We replace (2.2) by a.e. $G_{n,m}$ with $\delta \geq 2$ satisfies

(2.3) $G_{n,m} - A \in EX_2$ for all matchings A which avoid vertices of degree 2.

Observe also that a.e. $G_{n,m}$ is connected, has fewer than $\log n$ vertices of degree ≤ 2 and $\Delta \leq 3 \log n$. We finish the proof as before. Now define $\mathcal{G}_2(n,m)$ as those graphs in $\mathcal{G}(n,m)$ which are connected, non-hamiltonian and have $\delta \geq 2$, $\Delta \leq 3 \log n$ and fewer than log n vertices of degree ≤ 2. Now define $\mathcal{A}$ as the event

$$\{(i)\ A \text{ avoids at least one longest path of } G,\ (ii)\ A \text{ avoids vertices of degree 2},\ (iii)\ A \text{ is a matching.}\}$$

Then $\Pr(\mathcal{A} | G_{n,m} \in \mathcal{G}_2(n,m)) \geq \frac{1}{3}$ and $\Pr(\mathcal{A}) \leq (\frac{32}{33})^{\omega}$ by arguments similar to those for the previous theorem.

§3. Generalisations

Let a graph G have property $\mathcal{A}_k$ if it contains $\lfloor k/2 \rfloor$ edge disjoint Hamilton cycles and, if k is odd, a further edge disjoint matching of size $\lfloor n/2 \rfloor$, $n = |V(G)|$. Theorems 1.3, 1.4 can be generalised to

Theorem 3.1 (Bollobás and Frieze [1985]). $\lim_{n\to\infty} \Pr(\Gamma_{m_k^*} \in \mathcal{A}_k) = 1$ for fixed k.

This is not too difficult to prove. One shows that in a.e. $\Gamma_{m_k^*}$, is such that $\Gamma_{m_k^*} - A \in EX_k$ for all matchings A which avoid vertices of degree k. If then for example we can only find $t < \frac{1}{2}k$ Hamilton cycles then removing these cycles still leaves $|N(S)| \geq (k - 2t)|S| \geq 2|S|$ for $|S| \leq n/(k+2)$. Posa's Theorem plus the colouring argument finishes the proof.

It seems then that if we find that a random graph does not have property $\mathcal{A}_k$ then the most likely cause is a vertex of degree k-1 or less. But what if we exclude this possibility by only considering graphs with minimum degree at least k? Now let $\mathcal{G}^{(k)}_{n,m} = \{G \in \mathcal{G}_{n,m} : \delta(G) \geq k\}$ and let $G^{(k)}_{n,m}$ be sampled uniformly from this set. Bollobás and Frieze [1985] ($k = 1$) and Bollobás, Fenner and Frieze [1988] considered the probability that $G^{(k)}_{n,m} \in \mathcal{A}_k$. The questions to be answered are (i) how large should m be so that the probability tends to a positive constant and (ii) what is the most likely obstruction to the occurrence of $\mathcal{A}_k$? The answer to (ii) is the existence of $k+1$ vertices of degree k having a common neighbour - a *k-spider*. As for (i):

Theorem 3.2. Let $m = \frac{n}{2}\left(\frac{\log n}{k+1} + k \log n + c_n\right)$. Then

$$\lim_{n\to\infty} \Pr(G^{(k)}_{n,m} \in \mathcal{A}_k) = \lim_{n\to\infty} \Pr(G^{(k)}_{n,m} \text{ has no k-spider})$$

$$= \begin{cases} 0 & c_n \to -\infty \\ e^{-\theta(c,k)} & c_n \to c \\ 1 & c_n \to \infty \end{cases}$$

where $$\theta(c,k) = \frac{e^{-(k+1)c}}{(k+1)!((k-1)!)^{k+1}(k+1)^{k(k+1)}}.$$

[There is a caveat here for the case $c_n \to -\infty$. We should not allow it to go there too fast. For example if $k = 1$ and $m = \frac{1}{2}n$ exactly then $G^{(1)}_{n,m}$ is always a matching.]

The approach once again is to show that (in the key case, $c_n \to c$) a.e. $G^{(k)}_{n,m}$ is such that $G^{(k)}_{n,m} - A \in EX_k$ for matchings A that avoid vertices of degree k and to then apply the colouring argument. The first problem is quite difficult. It is hard to prove properties of $G^{(k)}_{n,m}$ since $\mathcal{G}^{(k)}_{n,m}$ is a very small subset of $\mathcal{G}_{n,m}$.

The k-core $\sigma_k(G)$ of a graph G is the largest vertex induced subgraph of G which has minimum degree at least k. Now it is not too difficult to show that if $\sigma_k(G_{n,m})$ has n' vertices and m' edges then it has the same distribution as $G^{(k)}_{n',m'}$. Now when m is as in Theorem 3.2, we find that in a.e. $G_{n,m}$ we have $n - n' = o(n)$, $m - m' = o(n)$ and hence we obtain the result of Luczak [1987a] as a corollary.

Corollary 3.3. Let m, $\theta(c,k)$ be as in Theorem 3.2. Then

$$\lim_{n\to\infty} \Pr(\sigma_k(G_{n,m}) \in \mathcal{A}_k) = \begin{cases} 0 & c_n \to -\infty \\ e^{-\theta(c,k)} & c_n \to c \\ 1 & c_n \to +\infty . \end{cases}$$

Note that the number of edges required in Theorem 3.2 for property $\mathcal{A}_k$ is much less than that required in Theorem 3.1.

The next thing to look at is $\mathcal{G}^{(k+1)}_{n,m}$. Here we find that only a linear number of edges is needed, although the result is not yet as precise as that in Theorem 3.2.

Theorem 3.4 (Bollobás, Cooper, Fenner and Frieze [1988]). There exist constants c_k, $k \geq 2$, such that if $m \geq c_k n$ then

$$\lim_{n\to\infty} \Pr(G^{(k+1)}_{n,m} \in \mathcal{A}_k) = 1.$$

Let us now for the moment turn to subgraphs of $G_{n,m}$. In particular let $m = \frac{1}{2}cn$ for some constant $c > 1$. Erdös conjectured that there exists a function $\alpha(c) > 0$, $\alpha(c) \to 0$ as $c \to \infty$ such that a.e. $G_{n,m}$ contains a path of length $\geq (1 - \alpha(c))n$. Ajtai, Komlós and Szemerédi [1981] proved this for all $c > 1$ and independently de la

Vega [1979] proved that $\alpha(c) \leq \frac{c_0}{c}$ for some small constant $c_0 > 1$. Ajtai, Komlós and Szemerédi showed that there was a long line of descendants in a certain branching process and de la Vega considered a simple depth-first-search for a long path. Bollobás [1982] was able to show that $\alpha(c)$ was much smaller, for large c, by showing that a.e. $G_{n,m}$ contained a large hamiltonian subgraph of size $n(1 - c^{24}e^{-c/2})$. The proof that the subgraph is Hamiltonian is based on Posa's Theorem plus the colouring argument. Looking for Hamiltonian subgraphs, at least for large c, seems to be the correct approach. Bollobás, Fenner and Frieze [1984] showed $\alpha(c) \leq c^6e^{-c}$ and in Frieze [1986a] we proved the following: let now k be fixed and

$$\alpha(k,c) = \inf\{\alpha \in \mathbb{R}: \text{ a.e. } G_{n,\frac{1}{2}cn} \text{ contains a subgraph } H \in \mathcal{A}_k \text{ with at least } (1-\alpha)n \text{ vertices}\}.$$

Theorem 3.5. $\alpha(k,c) \leq (1 + \varepsilon(k,c)) \sum_{t=1}^{k-1} \frac{c^te^{-c}}{t!}$ where $\lim_{c\to\infty} \varepsilon(k,c) = 0$.

This gives the correct order of magnitude for $\alpha(k,c)$ because a.e. $G_{n,\frac{1}{2}cn}$ contains approximately $n \sum_{t=1}^{k-1} \frac{c^te^{-c}}{t!}$ vertices of degree $k-1$ or less. In the proof we start with the k-core and remove a few extra vertices and show that what is left has property $\mathcal{A}_k$ with high probability. Theorem 3.5 is true if we allow c to grow with n and we can deduce Theorem 3.1 as a corollary. For the case c close to 1, Suen [1985] managed to improve the estimate of Ajtai, Komlós and Szemerédi for the length of the longest path.

Finally, Frieze and Jackson [1987] showed, at least for large enough c, that a.e. $G_{n,\frac{1}{2}cn}$ contains a vertex induced cycle of size at least $n/40\ c^3$.

Let us now return to the threshold for being Hamiltonian. A graph is said to be *pancyclic* if it has cycles of all lengths between 3 and n. Cooper and Frieze [1987] and Luczak [1987b] independently showed that the threshold for being pancyclic was also that for minimum degree 2. In fact Luczak showed that for large c, a.e. $G_{n,\frac{1}{2}cn}$ contains cycles of all lengths between $\omega(n)$ and the upper bound implied by Theorem 3.5. Here $\omega(n)$ is any function tending "slowly" to infinity with n. In a twist on pancyclicity Cooper [1988] has shown that when $G_{n,m}$ is Hamiltonian it is almost always possible to find a Hamilton cycle from which one can construct cycles of all lengths by using 2 chords for each cycle.

In another variation we showed, Frieze [1988a], that for any fixed k, the threshold for being Hamiltonian was also that for the existence of k vertex disjoint cycles of sizes $\lfloor\frac{n}{k}\rfloor$ and $\lceil\frac{n}{k}\rceil$ which covered [n].

Another variation on Hamiltonicity is that of being *Hamilton connected*. A graph is Hamilton connected if there is a Hamilton path joining each pair of vertices. To be Hamilton connected a graph has to have minimum degree at least 3 since the neighbours of a vertex of degree 2 cannot be connected by a Hamilton path. One can guess that the threshold for being Hamilton connected is the same as that for $\delta \geq 3$. This was proved in Bollobás, Fenner and Frieze [1987] and independently in Luczak [1988].

We ought to mention bipartite graphs. We showed in Frieze [1985], not surprisingly, that if $m = n(\log n + \log\log n + c_n)$, then

$$\lim_{n\to\infty} \Pr(B_{n,m} \text{ is Hamiltonian}) = \lim_{n\to\infty} \Pr(\delta(B_{n,m}) \geq 2)$$

$$= \begin{cases} 0 & c_n \to -\infty \\ e^{-2e^{-c}} & c_n \to c \\ 1 & c_n \to +\infty \end{cases}$$

where $B_{n,m}$ denotes a random bipartite graph with 2n vertices and m edges. The proof of this result was actually a little trickier than that of Theorem 1.2. The problem is that (2.1) is of no help if the longest path is of odd length. The trick used to deal with this was helpful in proving the next result.

We now describe a recent result of Cooper and Frieze [1988] on the number of Hamilton cycles in a random graph. Theorem 1.3 states that a.e. $\Gamma_{m_2^*}$ has a Hamilton cycle. This raises the question of how many? The expected number in $\Gamma_{m_2^*}$ is certainly at most $\frac{(n-1)!}{2}(\log n + O(\log\log n))^n = (\log n)^{n-o(n)}$. What we show is that a.e. $\Gamma_{m_2^*}$ has $(\log n)^{n-o(n)}$ distinct hamilton cycles. (The o(n) terms in the result and the expectation are quite different.) The idea behind the proof here is roughly to find for (most) $v \leq \frac{1}{2}n$, a collection of $k = \lceil\frac{c(\log n)^2}{r}\rceil$, (c constant, $r = (\log\log n)^2$), sets $W_1^{(v)}, W_2^{(v)}, \ldots, W_k^{(v)}$ of r edges satisfying $|W_i^{(v)} \cap W_j^{(v)}| \leq 1$, $i \neq j$. For each $f\colon [\frac{1}{2}n - o(n)] \to k$ we obtain a graph H_f by deleting all edges incident with v other than $W_{f(v)}^{(v)}$. We show that almost all H_f are Hamiltonian for a.e. $\Gamma_{m_2^*}$. If $f \neq f'$ then a Hamilton cycle in H_f is distinct from one in $H_{f'}$ since $|W_{f(v)}^{(v)} \cap W_{f'(v)}^{(v)}| \leq 1$ if $f(v) \neq f'(v)$. The number of different f is $k^{\frac{1}{2}n - o(n)} = (\log n)^{n-o(n)}$.

We end this section with a topic which we call *survival time*.

Suppose that the edges added in the graph process are $e_1, e_2, \ldots$ in this order and that the process stops when the graph is Hamiltonian. Suppose now that the graph starts to *deteriorate* and lose edges, oldest first. How long does the graph remain Hamiltonian? We discussed this question in Frieze [1988e]. We stated (without proof) that

$$\lim_{n\to\infty} P_r(\tau' \geq an) = e^{-2a} \quad , \quad a \in \mathbb{R}^+ ,$$

$$\lim_{n\to\infty} \frac{1}{n} E(\tau') = \frac{1}{2}$$

where τ' is the number of edges which are lost before the graph is no longer Hamiltonian. We called this the *survival time*.

§4. Regular Graphs, k-out and Planar Maps

There are two other graph models which have received some attention from the point of view of Hamiltonicity. We first consider random regular graphs. These are usually studied via the configuration model of Bollobás [1980]. Suppose r is constant and we wish to generate a random regular graph with vertex set $[n]$. We let $W = W_1 \cup W_2 \cup \ldots \cup W_n$ where the W_i's are a disjoint collection of r-sets. A configuration is a partition of W into $m = \frac{1}{2}rn$ pairs. Associate with F the multigraph $\mu(F)$ with vertex set $[n]$ and an edge xy whenever F contains a pair $\{\xi, \eta\}$ with $\xi \in W_x$, $\eta \in W_y$. If Φ is the set of configurations and F is chosen randomly from Φ then (i) each simple regular graph has the same probability of occurrence as $\phi(F)$ and

(ii) $\Pr(\phi(F)$ is a simple graph $\sim e^{-(r^2-1)/4}$. Hence if r is constant it is enough to show that a.e. $\phi(F)$ is Hamiltonian in order to show that a.e. r-regular graph is. This idea has been used to prove

Theorem 4.1. There exists a constant $r_0 \geq 3$ such that if $r \geq r_0$ is constant then a.e. r-regular graph is Hamiltonian.

Bollobás [1983] showed $r_0 \leq 10^{10}$ and independently Fenner and Frieze [1984] showed $r_0 \leq 776$. We gave an algorithmic proof that $r_0 \leq 85$ in [1988b]. Robinson and Wormald have claimed a proof that $r_0 = 3$, but as yet there is no paper. They have already proved that a.e. cubic bipartite graph is Hamiltonian and that 98% of cubic graphs are too [Robinson and Wormald, 1984]. They use the Chebycheff inequality to prove both results; the second result requires a clever argument relating cubic graphs and triangle free graphs. It would appear that the proof of $r_0 = 3$ is an extension of this argument.

One curious point about random regular graphs is that knowing a.e. r-regular graph is Hamiltonian does not imply a.e. (r+1)-regular graph is. Thus Theorem 4.1 does not say anything about r(n)-regular graphs when $r(n) \to \infty$. However by comparing rates at which various probabilities go to zero, one can show [Frieze, 1988c], that the result continues to hold for $r = O(n^{1/3-\epsilon})$. The upper bound seems unnecessary but is there for the moment.

There is not much to say about perfect matchings in this case since the existence question is covered by Hamiltonicity or r-connectivity.

Another model which has received some attention is $G_{k\text{-out}}$. Here the vertex set is [n] and each $v \in [n]$ independently chooses k neighbours. (Equivalently, sample uniformly from the space of digraphs with vertex set [n] and regular out degree k. Then ignore orientation). Observe that G_{1-out} is the graph induced by a random function and is well studied (see e.g. Kolchin [1986]). We proved the following in [Frieze, 1986b]:

Theorem 4.2. a.e. G_{2-out} has a matching of size $\lfloor n/2 \rfloor$.

This implies the earlier result of Shamir and Upfal [1982] the a.e. G_{6-out} has such a matching.

Now it is easy to see that a.e. G_{1-out} has no matching of this size since it will contain a large number of degree one vertices with a common neighbour. So how big is the largest matching in G_{1-out}. Let $\nu(n)$ denote its expected size.

Theorem 4.3 (Meir and Moon [1974]). $\frac{\nu(n)}{n} \to 1 - \rho$ where $\rho = .4328\ldots$ is the unique solution to $x = e^{-x}$.

Actually, Meir and Moon proved this result for random labelled trees. J.W. Moon pointed out to me that this implies the result for random mappings. This becomes obvious once one considers Joyal's proof of Cayley's formula for the number of spanning trees of K_n and the fact that a.e. G_{1-out} has $O(\sqrt{n})$ vertices on cycles. In the case of Hamilton cycles we have the following

Theorem 4.4. There exists a constant $k_0 \geq 3$ such that if $k \geq k_0$ then a.e. G_{k-out} is Hamiltonian.

It was shown in Fenner and Frieze [1983] that $k_0 \leq 23$ and it was here that we first used the counting argument used to finish the proofs of §2. We gave an algorithmic proof in [Frieze, 1988b] that $k_0 \leq 10$

but recently, in Luczak and Frieze [1988], we have shown $k_0 \leq 5$.

The idea of the last paper is simple enough to explain in a few paragraphs. G_{5-out} is (or contains) the union of 2 independent G_{2-out}'s plus a G_{1-out}. Take the 2 independent matchings from Theorem 4.2 and make up 2-factor which almost always has at most 2 log n cycles.

Now repeat the following procedure until a Hamilton cycle is constructed. Since a.e. G_{4-out} is connected there must be an edge joining 2 cycles. Delete 2 edges to make a path through all the vertices of these cycles. Now using only the G_{4-out} either extend the path to drag in another cycle and make the path longer, or do rotations and use the independent G_{1-out} to join 2 endpoints of one of the paths produced. Since G_{4-out} "expands" we can use Posa's Theorem to show there is always a high probability of being able to do this.

As a final model we consider planar maps and the result of Richmond, Robinson and Wormald [1985]. Consider the set of unlabelled 3-connected cubic planar maps with n vertices. They show that a.e. map contains a large number of vertex induced copies of any fixed size map M. By choosing M to be any 3-connected map which has no Hamilton path, they show that a.e. such map in non-Hamiltonian. Their proof is based on a clever use of generating functions.

§5. Algorithmic Aspects

The study of random graphs is by and large a study of the likely existence of objects. Random graphs can also be used to analyse the expected performance of algorithms that search for these objects. The Hamilton cycle problem is interesting from this point of view in that while the problem of finding a Hamilton cycle in a graph is NP-hard (see e.g. Garey and Johnson [1978]) there exist polynomial time algorithms which succeeds in almost every case. More precisely Bollobás, Fenner and Frieze [1987] devised a (deterministic) algorithm HAM which runs in $O(n^3 \log n)$ time on an n vertex graph and satisfies

Theorem 5.1. (i) Let $m = \frac{n}{2}(\log n + \log\log n + c_n)$. Assuming $c_n \to \pm\infty$ or a constant we have

$$\lim_{n\to\infty} \Pr(\text{HAM finds a Hamilton cycle in } G_{n,m})$$
$$= \lim_{n\to\infty} \Pr(G_{n,m} \text{ is Hamiltonian})$$

(ii) $\Pr(\text{HAM fails to find a Hamilton cycle in } G_{n,.5} \mid \delta(G_{n,.5}) \geq 2)$ $= o(2^{-n})$.

The main point of (ii) is that if the input to HAM is equally likely to be any graph with vertex set [n] then HAM fails so infrequently that these cases can be handled by dynamic programming (Held and Karp [1962]) and we will have a deterministic algorithm which correctly determines whether any graph is Hamiltonian and runs in polynomial expected time.

The algorithm uses the rotations discussed in §2. It runs in stages. At the start of stage k there is a path P of length k. If either endpoint of P is adjacent to a vertex not in P we extend the path and the stage ends. Otherwise we construct all paths obtainable by a single rotation and see if any of these can be extended. We continue this "breadth-first" construction of paths to depth approximately $T = \frac{2 \log n}{\log\log n}$, unless we find a path that we can extend. We show that in a.e. $G_{n,m}$, $m = \frac{n}{2}(\log n + \log\log n + c)$, the number of paths grows by a factor of at least $\alpha \log n$, $\alpha > 0$ constant. Thus at depth T we have approximately βn^2 paths and we use the colouring argument to show that with probability $1 - o(1)$ one of these paths has adjacent endpoints. Since a.e. $G_{n,m}$ will be connected we find that either we have a Hamilton cycle or we can find a longer path.

If we allow randomized algorithms then we have

Theorem 5.2 (Gurevich and Shelah [1987], Thomason [1987]). There exist linear expected time randomized algorithms for deciding graph Hamiltonicity for input distribution $G_{n,p}$, p constant.

Actually Thomason's paper treats $p \geq n^{-1/3}$. Neither of the two algorithms mentioned in this theorem have good expected performance at the threshold. On the other hand they can be modified to solve the Hamilton cycle problem on digraphs (see §6).

Earlier results for this problem were obtained by Angluin and Valiant [1979] who gave an $O(n(\log n)^2)$ randomized algorithm that finds a Hamilton cycle in a.e. $G_{n,Kn\log n}$ for large K. Shamir [1983] gave an algorithm for m slightly above the threshold.

Before moving on to sparse graphs we mention the result of Gimbel, Katz, Lesniak, Scheinermann and Weirman [1987]. It is well known that the graph $G + xy$ is Hamiltonian if and only if G is Hamiltonian whenever $d_G(x) + d_G(y) \geq |V(G)|$ (Bondy and Chvatal [1976]). By adding all such edges, we obtain $c(G)$. We can repeat this and compute $c^2(G), c^3(G), \ldots, c^*(G)$, the closure of G. Gimbel et al proved

Theorem 5.3. In the following p is a constant, $0 \leq p \leq 1$ and the statements hold with probability $1 - o(1)$:

(a) $p < \frac{1}{2}$ implies $c(G_{n,p}) = G_{n,p}$.

(b) $p = \frac{1}{2}$ implies $c^2(G_{n,\frac{1}{2}}) \neq c^3(G_{n,\frac{1}{2}}) = c^*(G_{n,\frac{1}{2}}) = K_n$.

(c) $p > \frac{1}{2}$ implies $c(G) = K_n$.

(Here $G_{n,p}$ is the random graph with vertex set [n] in which each possible edge occurs independently with probability p).

Let us now consider $G_{n,m}$, $m = \frac{1}{2}cn$ where $c > 1$ is constant. Karp and Sipser [1981] discuss a simple algorithm for finding a large matching in such a graph. Their algorithm is as follows: Suppose the input graph is G;

begin

H: = G - {isolated vertices}; M := ∅ ;

repeat

if $\delta(H) = 1$ **then** randomly choose an edge u incident with a vertex of degree 1 **else** randomly choose any edge u;

M := M + u; remove u and all edges indidcent with u; remove all isolated vertices;

until H has no vertices;

output M ;

end

They prove

Theorem 5.4. With probability $1 - o(1)$ the following hold.

(i) The matching produced by M is within o(n) in size of a largest matching.

(ii) If $c < e$ (= basis of natural logarithms) then for a.e. $G_{n,m}$ we find $\delta(H) = 1$ throughout the algorithm, otherwise there are phases in which $\delta(H) = 2$.

For the problem of finding a long path or cycle we essentially have de la Vega's algorithm which for large c finds a path within $O(\frac{n}{c})$ of the maximum and that in Frieze (1988b) which finds one within $O(n\ \epsilon(c)e^{-c})$ of the maximum, where $\lim_{c\to\infty} \epsilon(c) = 0$. Both algorithms are polynomial, the latter being a modification of HAM of Bollobás, Fenner and Frieze [1987]. This latter algorithm can also be used on random regular graphs.

Let us end this section with a discussion of parallel algorithms. Using the algorithm of Thomason [1987] to construct small cycles and then doing some patching we constructed an $O((\log\log n)^2)$ expected time parallel algorithm for deciding graph or digraph Hamiltonicity on a CRCW PRAM (Frieze [1987a]). The model of random input is $G_{n,p}$, p constant.

For sparse random graphs with $Kn \log n$ edges, K sufficiently large, Coppersmith, Rhagavan and Tompa [1987] have constructed an $O((\log n)^2)$ parallel algorithm that constructs a Hamilton cycle with probability $1 - o(1)$. Finally, in Frieze and Tygar [1988] we have been considering the problem of finding a maximum matching in a random graph. The algorithm is deterministic and purely graph theoretic in nature. It is to be contrasted with the *randomized* NC algorithms of Karp, Upfal and Wigderson [1986] or Mulmuley, Vazirani and Vazirani [1987] which work on <u>all</u> inputs and rely on evaluating determinants.

Weighted Problems

If we assign weights to the edgs of graphs then we can study the problems of finding minimum (total) weight perfect matchings and minimum weight Hamilton cycles (*the travelling salesman problem*). We first consider weighted perfect matchings in the complete bipartite graph $K_{n,n}$ (*the assignment problem*). The most efficient algorithms for this problem run in $O(n^3)$ worst-case time but Karp [1980] describes an algorithm with $O(n^2 \log n)$ expected running time assuming the weights are independent identically distributed random variables.

Suppose next that W_n is the random minimum weight of a perfect matching in $K_{n,n}$ when the edge weights are independent uniform [0,1] random variables. Walkup [1979] proved the surprising result that $E(W_n) \leq 3$, always. In the proof Walkup replaces each edge by a pair of edges, one blue and one red say. Each edge is given a weight which is a random variable with distribution function $F(x) = 1 - (1-x)^{1/3}$ so that the minimum of the red weight and the blue weight is uniform. Now consider the random bipartite graph where one half of the vertex partition chooses the two least weight red edges incident with each vertex and the other half uses blue edges. This graph is like a bipartite 2-out and Walkup [1980] shows it has a matching with probability $1 - o(1)$. (compare with Theorem 4.2). The expected length of each edge in this matching is at most $\frac{3}{n}$ and Walkup's result follows, modulo some technical tidying up.

Subsequently Karp [1987], showed $E(W_n) \leq 2$ (see also Dyer, Frieze and McDiarmid [1986]). It is known that $E(W_n) \geq 1 + e^{-1}$ and the exact limiting value of $E(W_n)$ is unknown.

Now consider the Travelling Salesman Problem (TSP). It was shown by Beardwood, Halton and Hammersley [1959] that if $X_1, X_2, \dots, X_n$ are independently chosen randomly from within the unit square $[0,1]^2$ then

$$\Pr\left(\frac{L_n}{\sqrt{n}} \to \beta\right) = 1$$

where L_n is the length of the shortest "tour" through the n points and β is constant whose precise value is not known. Karp [1977] in a very influential paper constructed an $O(n \log n)$ algorithm which computes a tour which, with probability $1 - o(1)$, is very close to optimum.

Karp also considered the asymmetric TSP. Let the arcs of the complete digraph be given independent uniform [0,1] lengths. Karp [1979] gave an $O(n^3)$ algorithm which, with probability $1 - o(1)$, computes a tour which is very close to optimum. Later Karp and Steele [1985] and Dyer and Frieze [1988] further improved these results. The basic idea is to compute in $O(n^3)$ time a minimum weight set of vertex disjoint directed cycles which together cover all vertices. [The optimisation problem here is, essentially, the assignment problem mentioned previously]. A tour is then obtained by "patching" together the cycles. Dyer and Frieze show that this can usually be done at an extra cost of $O(\frac{(\log n)^4}{n})$.

The above algorithms for the travelling salesman problem are approximation algorithms. They do not aim to solve the problem exactly. In Frieze [1987b] we consider the symmetric TSP where the costs are independent random integers in the range [0,B]. We described an $O(n^3 \log n)$ *randomized* algorithm TSPSOLVE which satisfies the following: assume $B = B(n) = o(\frac{n}{\log\log n})$. Then

(5.1) $$\lim_{n\to\infty} \Pr(\text{TSPSOLVE finds an optimum solution}) = 1.$$

The idea is to identify a set X_0 of "troublesome vertices" and construct a set of vertex disjoint paths $\mathcal{P}$ which contain X_0 as interior points. $\mathcal{P}$ is to have minimum total edge weight among all such sets of paths. Having done this we use zero length edges to construct a Hamilton cycle containing the eges of $\mathcal{P}$. The algorithm used for finding such a cycle is elated to that in Frieze (1988b).

§6. Digraphs

Many of the theorems on Hamilton cycles in random graphs have natural analogues in digraphs, most of which have not been proved yet. However the directed analogues of Theorems 1.2, 1.3, 2.1 have been proved in Frieze [1988d]. Thus for example if $D_{n,m}$ is a random digraph with vertex set $[n]$ and m edges chosen randomly from $[n]^2$ then we have

Theorem 6.1. Let $m = n \log n + c_n n$. Then

$$\lim_{n\to\infty} \Pr(D_{n,m} \text{ is Hamiltonian}) = \begin{cases} 0 & c_n \to -\infty \\ e^{-2e^{-c}} & c_n \to c \\ 1 & c_n \to +\infty . \end{cases}$$

The proof is via the analysis of an $O(n^{1.5})$ time algorithm.

There is not much else to say about random digraphs except that there is an analogous result to (5.1) for the asymmetric TSP with $B = O(\frac{n}{\log n})$.

We end this section with a remarkable inequality due to McDiarmid [1981] which gives a result close to Theorem 6.1. Let $D_{n,p}$ denote a random digraph in which each possible edge is independently included with probability p.

Theorem 6.2. $\Pr(D_{n,p} \text{ is Hamiltonian}) \geq \Pr(G_{n,p} \text{ is Hamiltonian})$.

Proof: Let $e_1, e_2, \ldots, e_N$ be any enumeration of the edges of the complete graph K_n. We consider a sequence of random digraphs $H_1,H_2,H_3,\ldots,H_N = D_{n,p}$.

To construct H_i we do the following: if $j < i$ and $e_j = uv$ then we independently add arc uv with probability p and arc vu with probability p. If $j \geq i$ then we either add both uv and vu with probability p or add neither. Thus H_1 is $G_{n,p}$ with each undirected edge replaced by a pair of oppositely oriented edges. Thus Pr(H_1 is Hamiltonian) = Pr($G_{n,p}$ is Hamiltonian). We show Pr(H_{i+1} is Hamiltonian) $\geq$ Pr(H_i is Hamiltonian), $1 \leq i < N$, and the theorem follows. In fact, let ω represent the outcome of our experiment with edges e_j, $j \neq i+1$. We have

(6.1) $\Pr(H_{i+1} \text{ is Hamiltonian} \mid \omega) \geq \Pr(H_i \text{ is Hamiltonian} \mid \omega)$,

remembering that edges e_j, $j \neq i+1$, are treated the same in H_i and H_{i+1}. Let D_ω denote the digraph with edges made from the outcomes of experiments with edges e_j, $j \neq i$. Now let $e_{i+1} = uv$. Then there are 5 cases:

(i) D_ω has a Hamilton cycle.

(ii) D_ω has no Hamilton cycle but has Hamilton paths from u to v and from v to u.

(iii) D_ω has no Hamilton cycle and no Hamilton path from u to v but has one from v to u.

(iv) D_ω has no Hamilton cycle and no Hamilton path from v to n but has one from u to v.

(v) D_ω has no Hamilton cycle and no Hamilton path from u to v or from v to u.

Consider the following table in which position $(k, i+\theta)$ gives

$\Pr(H_{i+\theta}$ is Hamiltonian | (case (k)):

	i	i+1
(i)	1	1
(ii)	p	$2p-p^2$
(iii)	p	p
(iv)	p	p
(v)	0	0

It is clear that we have proved (6.1) and the theorem. □

Observe that Theorems 1.2 and 6.2 imply the result in Theorem 6.1 for $c_n - \log\log n \to \infty$. (It is straightforward to translate such results for $D_{n,p}$ to $D_{n,m}$ - see Theorem II.2 of Bollobás [1985]).

§7. Open Problems

1. Can Theorem 3.1 be generalized to

$$\lim_{n\to\infty} \Pr(\Gamma_m \in \mathcal{A}_{\delta(\Gamma_m)},\ m = 1,2,\ldots,N) = 1?$$

2. Are the following true (see Theorem 3.4)?

   (a) $c > 1$ implies a.e. $G^{(2)}_{n,cn}$ has a matching of size $\lfloor n/2 \rfloor$.

   (b) $c \geq \frac{3}{2}$ implies a.e. $G^{(3)}_{n,cn}$ is Hamiltonian.

3. Determine $\epsilon(k,c)$ of Theorem 3.5 to within $o(1)$, (as $n \to \infty$).

4. Show that if $m = \frac{n}{2}(\log n + \log\log n + c)$ and $\delta(G_{n,m}) \geq 2$ then with probability tending to 1 either

   (a) ∃ a Hamilton cycle H such that cycles of all lengths can be obtained by adding 1 chord

or

(b) (a) is false.

(Cooper [1988] has shown that 2 chords are enough).

5. Determine the threshold for being able to partition the vertices of $G_{n,m}$ into $k = k(n)$ cycles of roughly equal size. (k constant is dealt with in Frieze [1988a]). The problem gets harder as k grows faster.

6. Show that a.e. $r(n)$-regular graph is Hamiltonian where $r(n) \to \infty$. The case $r = O(n^{1/3-\epsilon})$ is treated in Frieze [1988c] (I do not know at present whether showing that a.e. cubic graph is Hamiltonian is open).

7. Show that a.e. $G_{3\text{-out}}$ is Hamiltonian.

8. Find polynomial time algorithms for finding, with probability $1 - o(1)$, Hamilton cycles in (i) random cubic graphs, (ii) $G_{3\text{-out}}$.

9. Find polynomial time algorithms for solving random travelling salesman problems exactly, with probability $1 - o(1)$.

10. (a) Show that there exists a constant $r_0 \geq 2$ such that if $r \geq r_0$ then a.e. r-regular digraph is Hamiltonian. (By r-regular we mean both the indegree and outdegree of every vertex is r).
(b) Show that there exists a constant $k_0 \geq 2$ such that if $k \geq k_0$ then a.e. $G_{k\text{-in},k\text{-out}}$ is Hamiltonian. ($G_{k\text{-in},k\text{-out}}$ is a digraph with vertex set $[n]$ in which each $v \in [n]$ independently

chooses k in-neighbours and k out-neighbours. See Fenner and Frieze [1982]).

References

M. Ajtai, J. Komlos and E. Szemeredi (1981), "The longest path in a random graph" *Combinatorica* **1**, 1-12.

M. Ajtai, J. Komlos and E. Szemeredi (1985), "First occurrence of Hamilton cycles in random graphs", *Annals of Discrete Mathematics* **27**, 173-178.

D. Angluin and L.G. Valiant (1979), "Fast probabilistic algorithms for Hamilton circuits and matchings", *Journal of Computer and System Science* **18**, 155-193.

J. Beardwood, J.H. Halton and J.M. Hammersley (1959), "The shortest path through many points", *Proceedings of the Cambridge Philisophical Society* **55**, 299-327.

B. Bollobás (1980), "A probabilistic proof of an asymptotic formula for the number of labelled regular graphs", *European Journal on Combinatorics* **1**, 311-316.

B. Bollobás, (1982), "Long paths in sparse random graphs", *Combinatorica* **2**, 223-228.

B. Bollobás (1983), "Almost all regular graphs are Hamiltonian", *European Journal on Combinatorics* **4**, 97-106.

B. Bollobás (1984), "The evolution of sparse graphs", in *Graph Theory and Combinatorics, Proceedings of Cambridge Combinatorial Conference in Honour of Paul Erdös*, (B. Bollobás, Ed.) Academic Press, 35-57.

B. Bollobás (1985), *Random Graphs*, Academic Press.

B. Bollobás, T.I. Fenner and A.M. Frieze (1984), "Long cycles in sparse random graphs", in *Graph Theory and Combinatorics, Proceedings of Cambridge Combinatorial Conference in honour of Paul Erdös*, (B. Bollobás, Ed.) Academic Press, 59-64.

B. Bollobás, T.I. Fenner and A.M. Frieze (1987), "An algorithm for finding Hamilton paths and cycles in random graphs", *Combinatorica* **7**, 327-341.

B. Bollobás, T.I. Fenner and A.M. Frieze (1988), "Hamilton cycles in random graphs of minimal degree at least k", in preparation.

B. Bollobás, C. Cooper, T.I. Fenner and A.M. Frieze (1988), "Hamilton cycles in sparse random graphs of minimal degree at least k", in preparation.

B. Bollobás and A.M. Frieze (1985), "On matchings and Hamilton cycles in random graphs", *Annals of Discrete Mathematics* **28**, 23-46.

J.A. Bondy and V. Chvatal (1976), "A method in graph theory", *Discrete Mathematics* **15**, 111-136.

C. Cooper (1988), "Pancyclic Hamilton cycles in random graphs", to appear.

C. Cooper and A.M. Frieze (1987), "Pancyclic random graphs", to appear.

C. Cooper and A.M. Frieze (1988), "On the number of Hamilton cycles in a random graph", to appear.

D. Coppersmith, P. Raghavan and M. Tompa (1987), "Parallel graph algorithms that are efficient on average", *Proceedings of the 28th Annual IEEE Symposium on Foundations of Computer Science*, 260-269.

M.E. Dyer and A.M. Frieze (1988), "On patching algorithms for random asymmetric travelling salesman problems", to appear.

M.E. Dyer, A.M. Frieze and C.J.H. McDiarmid (1986), "On linear programs with random costs", *Mathematical Programming* **35**, 3-16.

P. Erdös and A. Renyi (1961), "On the evolution of random graphs", *Publ. Math. Inst. Hungar. Acad. Sci.* **5**, 17-61.

P. Erdös and A. Renyi (1966), "On the existence of a factor of degree of degree one of a connected random graph", *Acta. Math. Acad. Sci. Hungar.* **17**, 359-368.

T.I. Fenner and A.M. Frieze (1982), "On the connectivity of random m-orientable graphs and digraphs", *Combinatorica* **2**, 347-359.

T.I. Fenner and A.M. Frieze (1983), "On the existence of Hamilton cycles in a class of random graphs", *Discrete Mathematics* **45**, 301-205.

T.I. Fenner and A.M. Frieze (1984), "Hamilton cycles in random regular graphs", *Journal of Combinatorial Theory (B)* **37**, 103-112.

A.M. Frieze (1985), "Limit distribution for the existence of Hamilton cycles in a random bipartite graph", *European Journal on Combinatorics* **6**, 327-334.

A.M. Frieze (1986a), "On large matchings and cycles in sparse random graphs", *Discrete Mathematics* **59**, 243-256.

A.M. Frieze (1986b), "Maximum matchings in a class of random graphs", *Journal of Combinatorial Theory (B)* **40**, 196-212.

A.M. Frieze (1987a), "Parallel algorithms for finding Hamilton cycles in random graphs", *Information Processing Letters* **27**, 111-117.

A.M. Frieze (1987b), "On the exact solution of random symmetric travelling salesman problems with medium-sized integer costs", *SIAM Journal on Computing* **16**, 1052-1072.

A.M. Frieze (1988a), "Partitioning random graphs into large cycles", *Discrete Mathematics* **70**, 149-158.

A.M. Frieze (1988b), "Finding Hamilton cycles in sparse random graphs", *Journal of Combinatorial Theory (B)* **44**, 230-250.

A.M. Frieze (1988c), "On random regular graphs with non-constant degree", to appear.

A.M. Frieze (1988d), "An algorithm for finding Hamilton cycles in random directed graphs", *Journal of Algorithms* **9**, 181-204.

A.M. Frieze (1988e), "Survival time of a random graph", to appear in *Combinatorica*.

A.M. Frieze and B. Jackson (1987), "Large holes in sparse random graphs", *Combinatorica* **7**, 265-274.

A.M. Frieze and T. Luczak (1988), "Hamilton cycles in class of random graphs: one step further", to appear.

A.M. Frieze and D. Tygar (1988), "Deterministic parallel algorithms for matchings in random graphs", in preparation.

M.R. Garey and D.S. Johnson (1978), "Computers and Intractability: a Guide to the Theory of NP-Completeness, W.H. Freeman.

J. Gimbel, D. Kurtz, L. Lesniak, E.R. Scheinerman and J. Weirman (1987), "Hamiltonian closure in random graphs", *Annals of Discrete Mathematics* **33**, 59-67.

Y. Gurevich and S. Shelah (1987), "Expected computation time for Hamilton path problems", *SIAM Journal on Computing* **16**, 486-502.

M. Held and R.M. Karp (1962), "A dynamic programming approach to sequencing problems," SIAM *Journal on Applied Mathematics* **10**, 196-210.

R.M. Karp (1977), "Probabilistic analysis of partitioning algorithms for the travelling salesman in the plane", *Mathematics of Operations Research* **2**, 209-224.

R.M. Karp (1979), "A patching algorithm for the non-symmetric traveling salesman problem", *SIAM Journal of Computing* **8**, 561-573.

R.M. Karp (1987), "An upper bound on the expected cost of an optimal assignment", *Discrete Algorithms and Complexity: Proceedings of the Japan-US Joint Seminar*, (D. Johnson et al (eds.)), Academic Press, 1-4.

R.M. Karp and M. Sipser (1981), "Maximum matchings in sparse random graphs", *Proceedings of the 22nd Annual IEEE Symposium on Foundations of Computer Science*, 364-375.

R.M. Karp (1980), "An algorithm to solve the $m \times n$ assignment problem in expected time $O(mn\log n)$," *Networks* **10**, 143-152.

R.M. Karp and J.M. Steele (1985), "Probabilistic analysis of heuristics", in the *Traveling Salesman Problem: a Guided Tour*, (E.L. Lawler, J.K. Lenstra, A.H.G. Rinnooy-Kan and D.B. Shmoys (eds.)), John Wiley and Sons.

R.M. Karp, E. Upfal and A. Wigderson (1986), "Constructing a maximum matching is in random NC", *Combinatorica* **6**, 35–48.

V.F. Kolchin (1986), *Random Mappings*, Optimization Software Inc., Publications Division.

J. Komlós and E. Szemerédi (1983), "Limit distriubtions for the existence of Hamilton cycles in a random graph", *Discrete Mathematics* **43**, 55–63.

A.D. Korshunov (1976), "Solution of a problem of Erdös and Renyi on Hamilton circuits in non-oriented graphs", *Soviet Mathematics Doklaidy* **17**, 760–764.

T. Luczak (1987a), "On matchings and Hamiltonian cycles in subgraphs of random graphs", *Annals of Discrete Mathematics* **33**, 171–185.

T. Luczak (1987b), "Cycles in random graphs", to appear.

T. Luczak (1988), "On k-leaf connectivity of a random graph", *Journal of Graph Theory* **12**, 1-11.

C.J.H. McDiarmid (1981), "General percolation and random graphs", *Advances in Applied Probability* **13**, 40–60.

A. Meir and J.W. Moon (1974), "The expected node-independence number of random trees", *Nederl. Akad. Wetensch. Proc. Ser. Indag. Math.* **35**, 335–341.

K. Mulmuley, U.V. Vazirani and V.V. Vazirani (1987), "Matching is as easy as matrix inversion", *Proceedings of the 19th Annual ACM Symposium on Theory of Computing*, 345–354.

L. Posa (1976), "Hamilton circuits in random graphs", *Discrete Mathematics* **14**, 359–364.

L.B. Richmond, R.W. Robinson and N.C. Wormald (1985), "On Hamilton cycles in 3-connected cubic maps", *Annals of Discrete Mathematics* **27**, 141-150.

R.W. Robinson and N.C. Wormald (1984), "Existence of long cycles in random cubic graphs" in *Enumeration and Design* (D.M. Jackson and S.A. Vanstone, eds.) *Proceedings of Waterloo Conference on Combinatorics*, 251-270.

E. Shamir (1983), "How many random edges make a graph Hamiltonian?", *Combinatorica* **3**, 123–132.

E. Shamir and E. Upfal (1982), "One-factors in random graphs based on vertex choice", *Discrete Mathematics* **41**, 281–286.

S. Suen (1985), "Flows, cliques and paths in random graphs, Ph.D. Thesis, University of Bristol.

A. Thomason (1987), "A simple linear expected time algorithm for the Hamilton cycle problem", to appear.

W.T. Tutte (1947), "The factorization of linear graphs", *Journal of the London Mathematical Society* **22**, 107-111.

W.F. de la Vega (1979), "Long paths in random graphs", *Studia Sci. Math. Hungar.* **14**, 335-340.

D.W. Walkup (1979), "On the expected value of a random assignment problem", *SIAM Journal on Computing* **8**, 440-442.

D.W. Walkup (1980), "Matchings in random regular bipartite graphs", *Discrete Mathematics* **31**, 59-64.

Department of Mathematics
Carnegie Mellon University
Pittsburgh, PA 15213
U.S.A.

Decompositions of complete bipartite graphs
by
Roland Häggkvist

0. Setting the stage

Given two graphs G and H and an inquisitive mind we may ask whether or not G is the edge-disjoint union of copies of H. If the answer is yes we record this fact in short notation as $H|G$, read H divides (or decomposes) G and say that G admits an H-decomposition. (I shall also use the more active expression H packs G for the same notion). If not, an additional slash conveys *that* information. Thus for instance, using standard notation where kG denotes the graph consisting of k copies of the graph G, P_k the path with k vertices, C_k ditto cycle and K_n the complete graph on n vertices, etc., we have

0.1 $P_{2n}|K_{2n}$ (well-known)

0.2 $(C_4 \cup C_5) \not| K_9$ (Kotzig, see Guy(1971))

0.3 $2C_6 \not| K_{6,6}$ (easy exercise)

0.4 $6K_3 \not| K_{6,6,6}$ (Tarry 1900, see also proposition 1 below)

0.5 $nK_2|G$ G any k-regular graph on $2n \leq \frac{7}{6}k$ vertices (Chetwynd & Hilton (1983))

0.6 $K_{11} \not| K_{111}$ (Current sensation: The projective plane of order 10 does not exist! Announced by J. McKay, C.W.H. Lam, L.H. Thiel and S. Swiercs 1988. Research proposal outlined in (Lam et al.(1988))).

At this point I had better interject some remarks about the purpose of this paper. Well, to begin with it is not a survey. Instead it can be seen as a mixture between a serious research paper and an introductory discussion of a few standard decomposition problems for which I have developed, or more accurately put, stumbled on some easy bare hand techniques which nonetheless give uncommonly general answers. Part of the reason for mentioning these largely unexploited ideas lies in their simplicity and likely potential for generalisation; I have rarely pushed the approach to its limit.

The particular problems I shall discuss more in detail are (apart from the new theorems concerning decompositions of a complete bipartite graph into copies of some given regular graph) the twentyfive years old Ringel conjecture stating that every n-edge tree decomposes K_{2n+1} and the Oberwolfach problem, which here shall be thought of as the problem of determining the set of 2-factors which decompose a given $2k$-regular graph G (Recall that a 2-factor of G is a 2-regular spanning, i.e $|V(G)|$-order, subgraph of G). Both these problems have a large literature, almost none of which will be quoted. There are some glaring omissions. I shall not discuss hamilton decompositions or 1-factorizations, both topics that would fit very well with the rest of the paper, and I shall say little about design theory. In my view far too much effort is spent trying to implement the highly specialised design theoretic methods to other kinds of decomposition problems where often simpler ideas suffice.

The main new results in this paper are that every 3-regular bipartite graph B on $2n$ vertices with no component a Heawood graph (i.e the point-line incidence graph of the Fano-plane; the graph is explicitly shown in figure 1) decomposes $K_{6n,6n}$ (in fact $6B|K_{6n,6n}$), that every k-regular bipartite $2n$-order graph decomposes $K_{k^2!n,k^2!n}$ and therefore also $K_{2k^2!^2n^2+1}$, that every tree with n edges and having at least $\frac{n+1}{2}$ ends decomposes the complement of $(2n+1)C_3$, and a few results on packing lists of trees. Also a number of problems are given, usually in form of some general conjecture (it is definitely not hard to make plausible guesses in this area) which is unlikely to be answered in full, but where substantial special cases ought to be decidable.

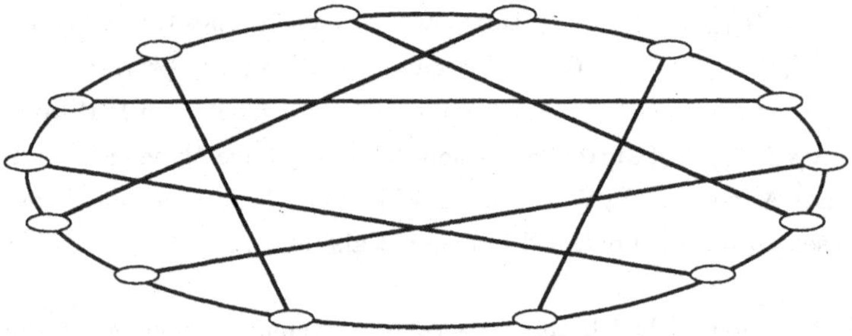

The Heawood graph
Figure 1.

After this digression let us continue where we were, with some introductory discussion of terminology and general background.

In this paper the symbol $\cup$ when used as a binary operation on unlabelled graphs means vertex-disjoint union (thus $C_6 \cup C_8$ has two components), and when used for labelled graphs where the graphs may have overlapping vertex sets $G_1 \cup G_2$

would denote the graph with vertex set $V(G_1) \cup V(G_2)$ and edge set $E(G_1) \cup E(G_2)$. The symbol $\oplus$ shall denote edge-disjoint union so that in particular if xy is an edge in both G_1 and G_2, say, then xy has multiplicity 2 (at least) in $G_1 \oplus G_2$. Moreover, a list $\mathcal{L} = (L_1, L_2, \ldots, L_q)$ of graphs is said to ***pack into*** a graph G if $G = G_1 \oplus G_2 \oplus \cdots \oplus G_q \oplus R$ where $G_i \simeq L_i$ for $i = 1, 2, \ldots, q$ and the list is said to decompose, or pack, G (again in short $\mathcal{L}|G$) if R is edge-less. In other words, $\mathcal{L}$ decomposes G if and only if there exists some colouring of the edges in G using q colours such that the ith colour induces a copy of L_i for $i = 1, 2, \ldots, q$.

Just two further bits of notation (terms not explicitly defined in this section or elsewhere are hopefully standard when not easily guessed at). Given the graph G we denote by $G^{(r)}$ the graph obtained by replacing each edge in G by r edges in parallel, and by $G(m)$ the graph obtained by replacing each vertex x by a set X of m new independent vertices $x^1, x^2, \ldots, x^m$ and for every pair of vertices x, y in $V(G)$ joining each vertex in X to each vertex in Y by altogether exactly $\mu(x, y)$ edges where $\mu(x, y)$ denotes the multiplicity (possibly 0) of the edges between x and y in G. In figure 2 is displayed a B-decomposition of $P_8(2)$ where B is the 2-regular graph consisting of one 8-cycle and one 6-cycle. Identifying the two ends of the path to create a 7-cycle, and then looking at figure 2 again we see that $(C_6 \cup C_8)|C_7(2)$.

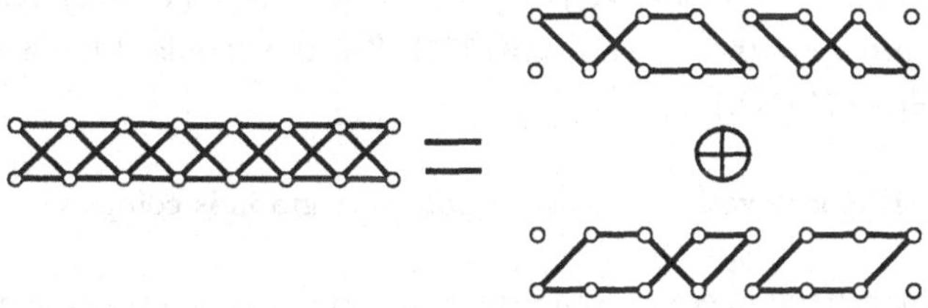

$(C_6 \cup C_8)|P_8(2)$

Figure 2.

It may be quite hard to decide even for specified G and H whether or not H divides G – the money and time spent to determine whether or not $K_{11}|K_{111}$ should make that remark selfevident if not downright foolish –, but as mathematicians rarely bother about special cases anyway, we may as well ask the following more general question.

The graph decomposition problem: *Give families $\mathcal{G}(n)$ of n-order graphs and families $\mathcal{H}(n)$ of suitable graphs of order at most n such that every member of $\mathcal{H}(n)$ decomposes every member of $\mathcal{G}(n)$.*

In fact we may often be even more optimistic and ask for decompositions into specified lists of graphs. For simplicity we use the term *proper list* for a graph G to mean that the entries in the list are graphs of order at most $|V(G)|$ and which have $|E(G)|$ edges in total, and so we have

The list-factorization problem: *Give families $\mathcal{G}(n)$ of n-order graphs and families $\mathcal{H}(n)$ of suitable graphs of order at most n such that every proper list $\mathcal{L} = (L_1, L_2, \ldots, L_q)$ with entries in $\mathcal{H}(n)$ decomposes every member G of $\mathcal{G}(n)$.*

The hundreds, not to say thousands, of papers dealing with the construction of balanced incomplete block designs testify to the interest in the problem of obtaining values (n, k, λ) for which there exist K_k-decompositions of $K_n^{(\lambda)}$. Worth remembering, and wellknown, are for instance the following facts.

0.7 $K_3 | K_n$ *if and only if* $n \equiv 1, 3$ mod.6 (Steiner triple systems (Kirkman 1847)),

0.8 $2nK_3 | K_{2n+1}(3)$ *for every* n, (Kirkman triple systems)

and

0.9 *For* $k = 3, 4$ *or* 5 *and every natural number* λ, *we have that* $K_k | K_n^{(\lambda)}$ *if and only if* $k - 1 | \lambda(n-1)$ *and* $\binom{k}{2} | \lambda\binom{n}{2}$ *unless* $(n, k, \lambda) = (15, 5, 2)$ *when no such decomposition exists.* (Hanani(1975). See also standard literature such as Street&Street(1987).).

However – a trite observation, but still – not every graph is complete.

The most general theorem concerning the graph decomposition problem to date is probably the one given by R.M Wilson (Wilson(1975)) at the Fifth British Combinatorial Conference in Aberdeen 1975.

Wilson's theorem (graph theory variant): *Every sufficiently large complete graph K_n is the edge-disjoint union of copies of the p-order graph H provided that*

i) $|E(H)|$ *divides* $\binom{n}{2}$ *and moreover*

ii) $GCD(d_1, d_2, \ldots, d_p) | (n-1)$ *where* d_i, $i = 1, 2 \ldots, p$ *are the degrees in* H.

Wilson's theorem is obtained using a fair amount of number theory and it would be of interest to have a more direct proof. It also remains to bring the constants implied in the expression "sufficiently large" down to earth. (At the moment they are, to quote Marshall Hall (Hall(1978)) in a short exposé: "though computable, still very large indeed".) My guess, for all that is worth, is that Wilson's theorem is true already when $n \geq p^2$, at least when H is regular, and that eventually someone shall find a proof to this effect.

Reading between the lines of a well worded problem posed by Nash-Williams 1970 [Nash-Williams(1970)] we arrive at a natural variation of Wilson's theorem.

Conjecture 0.10: *Every sufficiently large n-order graph G admits a K_p-decomposition if only*

a) every vertex has degree divisible by $p-1$,

b) the minimum degree is larger than $c_p n$ for some (smallest) constant $c_p < 1$ depending only on p

and

c) the number of edges in G is divisible by $\binom{p}{2}$.

This conjecture has recently been proved by T.Gustavsson (Gustavsson (1987)), a student of mine at the University of Stockholm, in the case when $p = 3$. He gets $c_3 < (1 - 10^{-24})$ so that in particular $n > 10^{24}$. Nash-Williams explicitly suggests that $c_3 = \frac{3}{4}$ might work if $n \geq 15$, say.

1. A small detour. Some connections with latin squares

The question we shall discuss, briefly, is the following: Given a graph G and a natural number m, when is it true that $mG|G(m)$? When that $G|G(m)$?

This is an old question when G is complete. In fact Leonhard Euler's 1782 paper on graeco-latin squares effectively showed that $mK_3|K_3(m)$ whenever m is odd or divisible by 4 and strongly suggested that $6K_3 \not| K_3(6)$, an odd fact which took some 120 years to prove (Tarry (1900), see [Denés & Keedwell(1974)]). The connection is explicitly written out in the following proposition.

Proposition 1.1: *The following three assertions are equivalent.*

i) $K_{k+1}|K_{k+1}(m)$

ii) $mK_k|K_k(m)$

iii) $N(m) \geq k-1$ *where* $N(m)$ *denotes the maximum number of pairwise orthogonal latin squares of order* m.

Proof: Let B be the $(k+1)$-partite graph $K_{k+1}(m)$ with parts $V_1, V_2, \ldots, V_{k+1}$, each on m vertices, so that $V_i = \{v_i^1, v_i^2, \ldots, v_i^m\}$ for $i = 1, 2, \ldots, k+1$. Let $A^1, A^2, \ldots A^{k-1}$ be any set of pairwise orthogonal $m \times m$ latin squares. Assume without loss of generality that A^j is based on the symbols $v_j^1, v_j^2, \ldots, v_j^m$ for $j = 1, 2, \ldots, k-1$. Put

$$V_{p,q} = \{v_j^i : v_j^i \text{ is the symbol in cell } (p,q) \text{ in } A^j\} \cup \{v_k^p, v_{k+1}^q\}$$

and let $B_{p,q}$ be the subgraph $B[V_{p,q}]$, i.e the subgraph of B induced by the vertices in $V_{p,q}$. Then clearly each $B_{p,q}$ is a copy of K_{k+1}, and moreover if $(a,b) \neq (p,q)$ then $B_{a,b}$ is edge-disjoint from $B_{p,q}$ since othervise if $x = v_i^j$, $y = v_r^s$, xy is an edge in both $B_{a,b}$ and $B_{p,q}$, and in addition $\{i,r\} \cap \{k, k+1\} = \emptyset$ then both x and y occur in the distinct cells (p,q) and (a,b) contradicting the orthogonality of A^i and A^r. If on the other hand $i = k$, say, then $a = p = j$ so that $x = v_k^p$. Consequently the symbol $y = v_r^s$ must occur in the two cells (p,b) and (p,q) in A^r contradicting the fact that A^r is a latin square. Similarly if $r = k+1$. Finally, since we have exactly m^2 edges between V_i and V_j in B for any $1 \leq i < j \leq m$ and since each of the m^2 graphs $B_{p,q}$ contains one of these edges all edges in B must belong to some $B_{p,q}$. We have therefore shown that B, a copy of $K_{k+1}(m)$, admits a K_{k+1}-decomposition given the particular sequence of mutually orthogonal latin squares. On the other hand we can invert the construction so that from any K_{k+1}-decomposition of B we obtain a sequence of pairwise orthogonal latin squares $A^1, A^2, \ldots, A^{k-1}$. Note that by relabelling the vertices of B we may end up with different latin squares, but nonetheless we have shown the equivalence between statements *iii)* and *i)*. Next, let $\tilde{B}_j$ be the union of the graphs $B_{i,j}$ for $i = 1, 2, \ldots m$. Then each $\tilde{B}_j$ consists of m edge-disjoint complete graphs sharing the vertex v^j_{k+1} so that $B_j = \tilde{B}_j - v^j_{k+1}$ is a copy of mK_k for $j = 1, 2, \ldots, m$. It follows that $B_1 \oplus B_2 \oplus \cdots \oplus B_m$ is a mK_k-decomposition of $B^* = B - V_{k+1}$, a copy of $K_k(m)$. Thus *i)* implies *ii)*. Finally, given B_j we can easily find $\tilde{B}_j$ and the graphs $B_{i,j}$ by joining the vertex v^j_{k+1} to all of B_j and letting $B_{i,j}$ be the clique in $\tilde{B}_j$ using the vertex v^i_k, for $i, j = 1, 2, \ldots, m$. Thus *ii)* implies *i)* and the proposition is proved. QED

We are now ready for the first theorem in this paper, an observation generalising proposition 1.1. A look at figure 3 at some stage during the proof is probably helpful, in particular to understand that part of the proof which is not explicitly written down (i.e the proof of 1.2 *i)*). In that figure is shown an explicit triangle-decomposition of $K_3(2)$ on the left and a derived C_5-decomposition of $C_5(2)$ on the right. (On the left the vertices in the columns labeled 3 should of course be identified so the graph is wrapped on a cylinder; similarly on the right. The labels indicate the colours in a proper vertex-3-colouring of K_3 and C_5 respectively.)

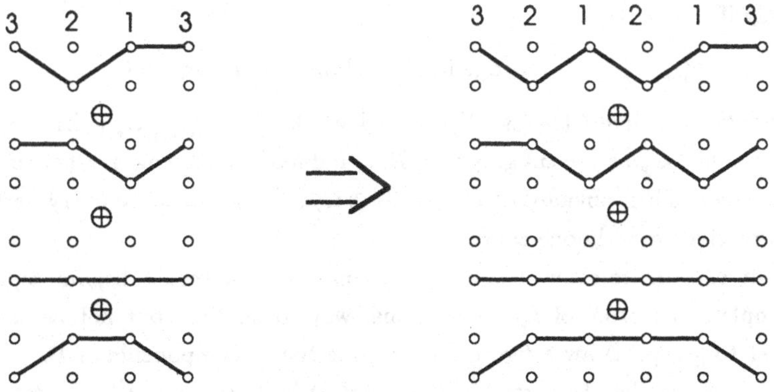

$K_3|K_3(2) \Rightarrow C_5|C_5(2)$

Figure 3.

Theorem 1.2: *Let G be a graph with chromatic number $\chi(G)$. Then, with $N(m)$ as in proposition 1.1 denoting the maximum number of pairwise orthogonal latin squares of order m we have*

i) $G|G(m)$ *if* $\chi(G) \leq N(m)+2$

and

ii) $mG|G(m)$ *if* $\chi(G) \leq N(m)+1$.

Proof: The proof is fairly straightforward using proposition 1.1, or rather perhaps the proof technique. Let us first prove *ii)*. Put $k = N(m)+1$ and as in the proof of proposition 1.1 we let B^* be a copy of $K_k(m)$, i.e a complete k-partite graph with equal sized parts $V_1, V_2, \ldots, V_k$ where $V_i = \{v_i^1, v_i^2, \ldots, v_i^m\}$ for $i = 1, 2, \ldots, k$. Proposition 1.1 guarantees that $B^* = B_1 \oplus B_2 \oplus \cdots B_m$ where each $B_i \simeq mK_k$. The decomposition of B^* induces in a natural way an edge-colouring $\lambda'_{i,j}$ of $B^*[V_i, V_j]$ (i.e the subgraph of B^* on $V_i \cup V_j$ containing the edges with one end in V_i and the other in V_j) where the edges of colour s, say are the edges in $B_s[V_i, V_j]$. Let $x_1, x_2, \ldots, x_n$ be the vertices in G. Then $V(G(m)) = \cup_{i=1}^n X_i$ where $X_i = \{x_i^1, x_i^2, \ldots x_i^m\}$ is the set of vertices stemming from x_i for $i = 1, 2, \ldots, n$. Finally, let $\lambda : V(G) \to \{1, 2, \ldots, k\}$ be a proper vertex k-colouring of G. Now let G_i, $i = 1, 2 \ldots, m$, be graphs on $V(G(m))$ defined by the following rule.

Rule : $$x_i^j x_r^s \text{ is an edge in } G_p$$

if and only if

$$v^j_{\lambda(x_i)} v^s_{\lambda(x_r)} \text{ is an edge in } B_p \text{ and } x_i x_r \text{ is an edge in } G.$$

In other words we give $G(m)[X_i, X_j]$ the edge-colouring $\lambda'_{\lambda(x_i),\lambda(x_j)}$ for $1 \leq i < j \leq n$ and define G_p as the subgraph of $G(m)$ induced by the edges of colour p for $p = 1, 2, \ldots, m$. Then obviously $G(m) = G_1 \oplus G_2 \oplus \cdots \oplus G_m$ since every vertex x in G has received exactly one colour $\lambda(x)$.

It only remains to show that every G_p consists of m vertex-disjoint copies of G to complete the proof of *ii)*. The formal way to do this (but not necessarily the easiest to grasp. Draw a figure!) is to label the m components of B_i as $B_{i,j}$, $j = 1, 2, \ldots, m$ and let the vertex in $V_r \cap V(B_{i,j})$ be the vertex $v_r^{f(i,j,r)}$, say (this defines a function $f : \{1, 2, \ldots, m\} \times \{1, 2, \ldots, m\} \times \{1, 2, \ldots, k\} \to \{1, 2, \ldots, m\}$). Note in particular that for fixed i and r the numbers $f(i,1,r), f(i,2,r), \ldots, f(i,m,r)$ are all distinct, since B_i consists of m components, all intersecting V_r in exactly one vertex. The graph $G_{i,j} = G(m)[A_{i,j}]$ where

$$A_{i,j} = \{x_s^{f(i,j,\lambda(s))} : s = 1, 2, \ldots, n\}$$

is isomorphic with G since $A_{i,j}$ contains exactly one vertex from each X_s, $s = 1, 2, \ldots, n$. Moreover $G_{i,j}$ has only edges of colour i since $x_r^{f(i,j,\lambda(r))} x_s^{f(i,j,\lambda(s))}$ is an edge in $G_{i,j}$ if and only if $v_{\lambda(r)}^{f(i,j,\lambda(r))} v_{\lambda(s)}^{f(i,j,\lambda(s))}$ is an edge in $B_{i,j}$. Now note that for fixed i and r, $r = \lambda(s)$, say, all vertices $x_s^{f(i,1,r)}, x_s^{f(i,2,r)}, \ldots, x_s^{f(i,m,r)}$ are distinct. Consequently all the graphs $G_{i,1}, G_{i,2}, \ldots, G_{i,m}$ are vertex-disjoint so that, finally, $G_i \simeq mG$ for $i = 1, 2, \ldots, m$. This proved *ii)*. The proof of *i)* is similar and left to the reader. QED

Remark: The above proof shows more. In fact, let us define a *faithful mG-decomposition of $G(m)$* as an colouring of the edges of $G(m)$ in m colours such that for each colour i the edges of colour i induce a copy G_i of mG for which in addition $E(G_i[X_r, X_s])$ is an m-matching whose edges come from distinct copies of G in mG for every edge $x_r x_s$ in G. Similarly, let us define a *faithful G-decomposition of $G(m)$* as an edge-colouring of $G(m)$ in m^2 colours such that for each i, $i = 1, 2, \ldots, m^2$ the graph G_i spanned by the edges of colour i form a copy of G for which moreover $E(G_i[X_r, X_s])$ has exactly one edge for every edge $x_r x_s$ in G. Finally recall that a graph G is homomorphic with a graph H if there exists some function $\tau : V(G) \to V(H)$ for which $\tau(x)\tau(y)$ is an edge in H for every edge xy in G. Thus $\chi(G) \leq k$ if and only if G is homomorphic with K_k. It should now be clear that the above proof actually works to give the following theorem.

Theorem 1.3: *Let G be a graph homomorphic with the graph H. Then the statements i) and ii) below are true.*

i) If $H(m)$ admits a faithful mH-decomposition, then $G(m)$ admits a faithful mG-decomposition,

and

ii) if $H(m)$ admits a faithful H-decomposition, then $G(m)$ admits a faithful G-decomposition.

2. The Ringel conjecture

In 1963 Gerhard Ringel proposed the folloving problem (no.25 in the problem session at the conference in Smolenice (Ringel (1963))).

Conjecture 2.1: *Show that every $(n+1)$-order tree decomposes K_{2n+1}.*

Rather, to be precise he stated: "It is conjectured that the complete $(2n+1)$-gon can be decomposed into $2n+1$ subgraphs which are all isomorphic to a given tree with n edges." The Ringel conjecture is often confused with the graceful tree conjecture which states that every tree admits a graceful labelling (Recall that a graceful labelling of a m-edge-graph H is an injection $\tau : V(H) \to \{0,1,\ldots,m\}$ such that $\{|\tau(x)-\tau(y)| : xy \in E(H)\} = \{1,2,\ldots,m\}$), a conjecture named and popularized by Golomb (see MR 49 # 4863), but probably originating from A. Kotzig (this is explicitly claimed in [Huang et al.(1982)]) as a means of attacking 2.1. In my view the graceful tree conjecture is something of a red herring and I fully agree with the exasperated words of Richard K. Guy: "Far too many (over a hundred) papers have appeared on the subject." (Quoted from MR 82g:05041 (1982)).

However, one consequence of trees being graceful would be that $2T|K_n(2)$ for every n-order tree T and that would certainly be a good enough theorem on its own, if true. In fact there is reason to believe that there exists some constant $c_k > 0$ such that every m-regular n-order graph G with $m > (1-c_k)n$ has the property that $kT|G(k)$ for every $(m+1)$-order tree T and every natural even number $k \geq 2$.

The following strengthening of conjecture 2.1, also attributed to Ringel, appears in (Guy(1971))

Conjecture 2.2: *Show that every tree on $n+1$ vertices decomposes K_{rn+1} for every natural number $r \geq 2$ provided that r and $n+1$ are not both odd.*

It is somewhat surprising that such a nice problem has remained open for so long. Surely the case $r \geq 10^{10}$, say, ought to be doable. To justify this remark slightly, let me show how to decompose $K_{2n+1}(3)$, a graph not too unlike K_{6n+1}, into copies of a given $(n+1)$-order tree with more than $\frac{n+1}{2}$ ends (a tree without vertices of degree 2 for instance). The proof technique, which is very general and which can be applied to a number of tree-decomposition problems, is explained in the following tale describing ...

...how to pack a porcupine.

Now before one packs a porcupine it helps to know what it is. Here is the description. A *porcupine* is a graph H with (usually many) vertices, called *quills*, of degree 1. The quills are nasty, so we may want to delete them; if we do that what remains is the body H'.

Suppose now that we want to decompose some $2m$-regular n-order graph G into n copies of the m-edge porcupine H and assume that we can pack the bodies uniformly into G in the sense that each vertex in H' is mapped onto each vertex of G exactly once. In other words we assume that we have coloured the edges of G using $n+1$ colours and such that each of the colours $1, 2, \ldots, n$ induces a copy of H' and moreover such that if x^i denotes the vertex x in H' when viewed in the i-coloured copy of H' then $\{x^i : i = 1, 2, \ldots, n\} = V(G)$ for every $x \in V(H')$. We are then left with the problem of embedding the quills into the $n+1$-coloured graph R in a proper fashion. This means that we want to give R an edge-colouring using n colours such that for $i = 1, 2, \ldots, n$

2.3 *the i-th colour class consists of a union of stars where the ends of one star do not overlap the ends of another (but the ends of one star could a priori overlap the center of some other star; this is ruled out in the next paragraph), centered at the vertices* $\{x^i : x \in V(H')\}$, *and such that the star centered at* x^i *has degree exactly* $d(x, H) - d(x, H')$.

and

2.4 *no edge of colour* i *in* R *shall join two vertices of the form* x^i, y^i, *for any pair of distinct vertices* $x, y \in V(H')$.

It is clear that if such an edge-colouring of R exists then G admits an H-decomposition, since the subgraph of colour i in $G' = G - E(R)$ together with the subgraph of colour i in R forms a copy of H for $i = 1, 2, \ldots, n$.

It now turns out that property 2.3 can always be achieved, whereas property 2.4 is harder to guarantee and requires some ad hoc argument. The reason is that 2.3 can be read out of the following edge-colouring theorem of mine (Häggkvist(1982)) which may look a bit out of place here, but which nonetheless does the trick, as we shall see.

First a definition. Let $B = B(X, Y)$ be a bipartite graph (here note that multiple edges are allowed.), where each vertex in X is assigned a set $f(x)$ of $d(x)$ colours from some set Z, say, of colours. We say that B admits an f-edge-coluring if the edges in B can be coloured using colours from Z such that every colour class is a matching and where in addition every edge xy, $x \in X$, has received a colour from $f(x)$.

Theorem 2.5: *Let $B = B(X,Y)$ be a bipartite graph where every vertex x in X is assigned a set $f(x)$ of $d(x)$ colours from a set of colours Z. Assume furthermore that B admits a proper edge-colouring in colours $1,2,\ldots,$ such that every vertex in X of degree at least i is incident with some edge of colour i, for $i = 1,2,\ldots.$ Then B admits an f-edge-colouring.*

Proof: The theorem follows immediately from the following observation. Let s and t be two elements of Z, and let $g : X \to 2^Z$ be given by

$$g(x) = \begin{cases} f(x) \setminus s \cup \{t\} & \text{if } s \in f(x) \text{ and } t \notin f(x) \\ f(x) & \text{else.} \end{cases}$$

In this situation we say that g is obtained from f by an elementary compression. Then we have

Lemma 2.6: *With f, g as above B admits an f-edge-colouring if it has an g-edge-colouring.*

To see this it suffices to consider the subgraph $B_{s,t}$ of B induced by the edges of colours s or t in B under a given g-edge-colouring. Note that every component in $B_{s,t}$ is an alternating path or cycle. The paths in $B_{s,t}$ have the property that every vertex in X which is incident with an edge of colour s must be incident with an edge of colour t as well, by the construction of g. Consequently all paths in $B_{s,t}$ which start at a vertex x in X must start with an edge of colour t and must end with some edge of colour t incident with some vertex y in Y (this argument used the fact that every path from X to X has even length since B is bipartite. In other words, it is not possible for an XX-path in $B_{s,t}$ to begin and end with the same colour and consequently there exists no XX-path in $B_{s,t}$.) If we now interchange colours on the paths in $B_{s,t}$ incident with the vertices x' in X for which $g(x') \neq f(x')$ we get an f-edge-colouring of B, and the lemma is proved.

The theorem follows since we can obtain the function $h(x) = \{1,2,\ldots,d(x)\}$, $x \in X$ from f by repeated elementary compressions. QED

One case where the assumption in theorem 2.5 is trivially fulfilled is if B is regular of degree m, say, since then B admits a proper m-edge-colouring by a well known theorem (König 1916). Another case is if all vertices in X except one have degree at least m while all vertices in Y have degree at most m (Easy exercise. Note that we can always match the set of largest vertices in X into Y. By doing this recursively until every vertex in X except possibly the smallest has degree m

and then applying König's theorem the assertion follows.). We therefore have the following corollary.

Corollary 2.7: *Let $B = B(X,Y)$ be a bipartite graph where every vertex in X except possibly one has degree at least m and every vertex in Y has degree at most m. Assume furthermore that every vertex x in X has been assigned a set $f(x)$ of $d(x)$ colours from Z. Then B admits an f-edge-colouring.*

Changing the role of Y and Z we get immediately a formulation where the colour classes no longer are matchings but instead forests of stars centered at X. The condition that every vertex in Y be of degree at most m translates to the condition that the each colour class has at most m edges, and the requirement that every vertex in x is assigned a *set* of colours from Z gives that we can only consider simple graphs.

The precise formulation is given now.

Corollary 2.8: *Let B be a simple* (sic!) *bipartite graph with bipartition (X,Z) where every vertex in X except at most one has degree at least m. Assign to each vertex x in X a multiset $h(x)$ of colours chosen from some t-set Y such that $|h(x)| = d(x)$. Assume furthermore that every colour from Y is used at most m times, counting multiplicity. Then B admits an h-edge-colouring in the sense that every colour induces a forest of stars where in addition the multiset $h(x)$ equals the multiset of colours on the edges incident to x, for every vertex x in X.*

Finally, by orientating all edges from X to Z and identifying every vertex in X with distinct vertices in Z (if necessary enlarging Z) to obtain a digraph D and reformulating the condition that every vertex in X is assigned a multiset to the equivalent condition that every vertex in $V(D)$ is assigned the incidence vector of a multiset we get theorem 2.9.

Theorem 2.9: *Let D be a simple digraph where all except possibly one vertex have out-degree at least m. Assign to each vertex v in D a t-dimensional vector $f(v) = (f_1(v), f_2(v), \ldots, f_t(v))$, each component a non-negative integer, such that $\sum_i f_i(v) = d^+(v)$ for every vertex v and moreover*

$$\sum_{v \in V(D)} f_i(v) \leq m \qquad \text{for } i = 1, 2, \ldots, t.$$

Then $D = D_1 \oplus D_2 \oplus \cdots \oplus D_t$ such that for every $v \in V(D)$ and every i, $i = 1, 2, \ldots, t$ we have $d^+(v, D_i) = f_i(v)$ and $d^-(v, D_i) \leq 1$.

There is really not much left of the porcupine story. It should be added perhaps that 2.3 can be fulfilled since we may orient R to produce an eulerian digraph which acts as the digraph D in theorem 2.9. The vector $f(v)$ in theorem 2.9 comes from the requirements on the degrees of the ith colour class in R (recall that we want to endow R with an edge-colouring in n colours such that all vertices of the form x^i have degree $d(x,H)-d(x,H')$), the ith component in $f(v)$ is either 0, if the vertex v in $V(D)$ is not x^i for any i, otherwise it is $d(x,H)-d(x,H')$ for the unique vertex x in $V(H')$ for which $v = x^i$.

After this dry run we are ready for the enunciated theorem.

Theorem 2.10: *Let T be a tree with n edges and with at least $\frac{n+1}{2}$ ends. Then $T|K_{2n+1}(3)$.*

Proof: The proof which uses proposition 1.1 and theorem 2.9 shall be given with a few of the standard details glossed over, in order to make the ideas somewhat clearer.

We proceed as follows. Delete the ends from T and call the resulting tree T'. Use the greedy algorithm to give a labelling to the vertices of T' using as labels the numbers $1, 2, \ldots, 2n+1$ such that no pair of edges have the same differences of edge-labels modulo $2n+1$. Use this labelling to pack $2n+1$ copies of T' cyclically into K_{2n+1} in a standard fashion. In other words, we let G be a complete graph on the $2n+1$ vertices $y_1, y_2, \ldots, y_{2n+1}$, say, and give G an edge-colouring using $2n+2$ colours such that the ith colour induces a copy of T', for $i = 1, 2, \ldots, 2n+1$, and moreover such that if the vertices in G are placed in cyclical order equally spaced along the perimeter of a circle with the edges of one of the $2n+1$ copies of T' drawn, then the other copies are gotten by rotating the circle. It is obvious that this edge-colouring is uniform in the sense that if the vertex x in T' occurs as the corresponding vertex x^i in the ith coloured copy of T' in G then $\{x^i : i = 1, 2, \ldots, 2n+1\}$ is the whole vertex set of G and this happens for every $x \in V(T')$. Let H be the subgraph of G spanned by the edges which do not belong to any of the copies of T'. We may note here in passing that H is regular of degree at least n, since by assumption T had at least $\frac{n+1}{2}$ ends so that $2n+1$ copies of T' must have at most $(2n+1)(n - \lceil\frac{n+1}{2}\rceil) \leq (2n+1)\frac{n}{2}$ edges. This, together with the particular structure of H (for instance the fact that H is a circulant) makes it probable that there exists some colouring of the edges of H in $2n+1$ colours where the ith colour class is a forest of stars centered at x^i, $x \in V(T')$, the star at x^i having $d(x,T) - d(x,T')$ edges, and where moreover no edge of colour i joins an x^i to an y^i, $x, y \in V(T')$, and then a T-decomposition of G would be achieved. However, as it is, we must make do with less.

First of all, note that H is regular of even degree (it is the complement of a circulant graph with an odd number of vertices). Consequently H can be oriented to produce a digraph D for which the in-degree equals the out-degree at every vertex. A brief calculation based on the fact that T has n edges shows that the out-degree at each vertex in D equals

$$\sum_{x \in V(T')} d(x,T) - d(x,T').$$

Moreover every vertex y in D has the property that for every $x \in V(T')$ there exists a unique $i = f(y,x)$ such that $y = x^i$. It follows that we can use theorem 2.9 to obtain an edge-colouring of D using $2n+1$ colours such that the edges in the ith colour class form a digraph D_i for which the out-degree at every vertex of the form x^i with $x \in V(T')$ equals the number $d(x,T) - d(x,T')$ and such that the in-degree at every vertex is at most 1. This latter fact implies that each component in the underlying graph H_i of D_i has at most one cycle (easy exercise).

Now, for every i, $i = 1, 2, \ldots, 2n+1$ we let G_i be the underlying graph of $\vec{G}_i$, the mixed graph spanned by the edges in the ith copy of T' and D_i. Then,

$$G = G_1 \oplus G_2 \oplus \cdots \oplus G_{2n+1}$$

so that theorem 2.10 follows if we can show that every G_i, $i = 1, 2, \ldots, 2n+1$, has the property that $G_i(3)$ admits a T-decomposition. This is done next, and in this step we shall once again use proposition 1.1, or rather, parts of the proof of theorem 1.2. However, we shall first establish that

2.11 $$\chi(G_i) \leq 4.$$

This follows from the fact that since G_i is the edge-disjoint union of a tree and a graph all of whose components have at most one cycle we must have that $|E(G_i[A])| \leq 2|A| - 1$ for every set set $A \subset V(G)$. Consequently G_i is 3-degenerate (i.e every subgraph has a vertex of degree at most 3) whence 2.11 is true.

It follows from theorem 1.2 that $G_i | G_i(3)$. Moreover, with Y_j denoting the set of vertices $\{y_j^1, y_j^2, y_j^3\}$ in $G_i(3)$ stemming from the vertex y_j in G_i it follows from the proof of theorem 1.2 that $G_i(3)$ admits an edge-colouring λ' in 9 colours such that each colour class uses exactly one edge of the 9 edges in $G_i[Y_k, Y_l]$ for every pair of adjacent vertices y_k and y_l in G_i and such that each colour class is a copy of G_i. Let $G_{i,1}, G_{i,2}, \ldots, G_{i,9}$ be a listing of the 9 copies of G_i in question, and let $y_k y_l$ be an edge in G_i. We endow $G_i[Y_k, Y_l]$ with an edge-colouring λ'' in 9 colours as follows. Let $y_k^r y_l^s$ be an edge of colour t, say in the edge-colouring λ', $t = 1, 2, \ldots, 9$. If $y_k y_l$ is an undirected edge in the mixed graph $\vec{G}_i$ we let $y_k^r y_l^s$ have colour t in λ'', otherwise if $y_k y_l$ is a directed edge in $\vec{G}_i$, say directed from y_k to y_l, then we let the edge $y_k^r y_l^{s+1}$ have colour t in λ'' (superscripts counted modulo 3). Do this for all k, l, r, s and t. The resulting edge-colouring λ'' has the property that each colour class of edges induces a copy of T. The verification is left to the reader.QED

It is a trend in modern mathematics that if you can not solve a particular problem you can at least formulate a more general one, thus increasing the amount of frustration in the world and keeping your collegues on their toes. This is partly the motivation behind the next conjecture, proposed by Ron Graham and myself some five years ago. It is also a natural generalisation of 2.1.

Conjecture 2.12: *Every $m+1$-order tree decomposes every $2m$-regular graph.*

In fact, for bipartite graphs there is a much stronger variant.

Conjecture 2.13: *Every $m+1$-order tree decomposes every m-regular bipartite graph.*

Some special cases of these conjectures in particular 2.14–2.16 below have been proved by me in a manuscript written five years ago, but never really polished off for publication. The proofs use the packed porcupine technique.

Theorem 2.14: *Every $(m+1)$-order tree T with at least $\frac{m+1}{2}$ end-vertices (for instance every tree without vertices of degree 2) decomposes $C_k(m)$ for $k = 3, 4, \ldots$. Thus in particular T decomposes $K_{m,m,m}$ and $K_{2m,2m}$.*

Theorem 2.15: *Every $(m+1)$-order tree with diameter at most k decomposes every $2m$-regular graph of girth at least k, for $k = 3, 4, \ldots$.*

Theorem 2.16: *Every $(m+1)$-order tree with diameter at most $2k$ decomposes every bipartite m-regular graph of girth at least $2k$ for $k = 2, 3, \ldots$.*

Finally, let me mention a still stronger statement which *may* be true, although prudence compels me to add that it seems somewhat too general to be provable in the near future.

Conjecture 2.17: *Let $\mathcal{L} = (L_1, L_2, \ldots, L_n)$ be a list of $(m+1)$-order trees, and let G be a $2m$-regular n-order graph. Then $\mathcal{L}|G$.*

The immediate strengthening for bipartite graphs is false as is seen e.g. by the fact that no 4-regular bipartite graph is the edge-disjoint union of one path of length 4 and 4-stars, and it is not at all unlikely that some equally trivial counterexample to 2.17 exists. However, 2.17 is true if G admits a hamilton decomposition provided that all trees in the list have diameter 3 at most, as shall be proved in theorem 2.18 below. It is also true in some other instances such as when the graph $G = K_{2m,2m}$, m is even and furthermore the list contains trees with vertices of degrees 1 and k, $k \geq 5$, only, but that shall be proved elsewhere.

Theorem 2.18: *Let G be a $2m$-regular n-order graph with a hamilton decomposition and let $\mathcal{L} = (L_1, L_2, \ldots, L_n)$ be a list of $(m+1)$-order trees of diameter at most 3. Then $\mathcal{L}|G$.*

Proof: Each L_i, $i = 1, 2, \ldots, n$ is a star or double star. Consequently there exists an edge $v_i w_i$ whose vertices are adjacent to each end-vertex in L_i. If necessary we reorder the list such that $d(v_i) \leq d(v_{i+1})$ for $i = 1, 2, \ldots, n-1$. We may also assume that $d(v_i) \leq d(w_i)$ for every i, but in what follows that fact shall not be used. Let $H_1 \oplus H_2 \oplus \cdots \oplus H_m$ be one of the postulated hamilton decompositions of G and assume that $x_1 x_2 \cdots x_n x_1$ is the hamilton cycle H_1. Pack the edges $v_i w_i$ into H_1 by the rule that w_n and v_1 map onto x_n while for $i = 1, 2, \ldots, n-1$, w_i and v_{i+1} map onto x_i. It is not to hard to see (the complete argument follows at the end of the proof so as not to slow down the discussion too much) that the graph $G - E(H_1)$ can be oriented such that the resulting digraph D fulfils $d^+(x_i) = d(v_{i+1}) + d(w_i) - 2$ for $i = 1, 2, \ldots, n-1$. Assuming this we note that the sequence $d^+(x_i)$, $i = 1, 2, \ldots, n-1$ consists of numbers greater than or equal to $m-1$ while $d(v_i) + d(w_i) = m+1$ for every i. We can therefore use theorem 2.9 to find a decomposition of D as $D_1 \oplus D_2 \oplus \cdots \oplus D_n$ where each digraph D_i consists of two stars centered at x_{i-1} and x_i respectively (subscripts counted modulo n), the first having $d(v_i) - 1$ edges while the other has $d(w_i) - 1$ ditto. This means that the edge $x_{i-1} x_i$ together with the underlying graph of D_i forms a copy G_i of L_i for $i = 1, 2, \ldots, n$. Consequently the proof is complete except for the orientation of $G - E(H_1)$ and that is done now, using a bit of formalism.

First of all notice that $d(v_{i+1}) + d(w_i) - 2 \geq d(v_i) + d(w_i) - 2 = m - 1$ for $i = 1, 2, \ldots, n-1$. Moreover using the fact that the number of edges in $G - E(H_1)$

is $n(m-1)$ and that $\sum_i(d(v_i)+d(w_i)) = n(m+1)$ we get

$$\sum_{i=1}^{n-1}(d(v_{i+1})+d(w_i)-2-m+1) = m+1-d(v_1)-d(w_n) \le m-1$$

Put $i_0 = 0$ and let $i_1, i_2, \ldots, i_p$ be the indices r for which $d(v_{r+1})+d(w_r)-m-1 > 0$. Moreover put $f(0) = 1$ and

$$f(j) = \sum_{i=1}^{j}(d(v_{i+1})+d(w_i)-m-1) \qquad \text{for } j = 1, 2, \ldots, n-1$$

Then $f(j)$, $j = 1, 2, \ldots, n-1$ is an increasing nonnegative integer-valued function and $f(n-1) \le m-1$. Moreover, $f(i_{k+1})-f(i_k) = d(v_{i_{k+1}+1})+d(w_{i_{k+1}})-m-1$ for $k = 0, 1, \ldots, p-1$ Now orient each of $H_{f(i_k)+1}, H_{f(i_k)+2}, \ldots, H_{f(i_{k+1})}$ so that $x_{i_{k+1}}$ is joined towards x_n by two directed paths for $k = 0, 1, \ldots, p-1$, and orient each of $H_{f(i_p)+1}, H_{f(i_p)+2}, \ldots, H_m$ cyclically. This orientation of $G - E(H_1)$ gives the digraph D (note that in particular $d^+(x_j) = m-1 = d(v_{j+1})+d(w_j)-2$ for $j \ne i_0, i_1, \ldots, i_p$ while $d^+(x_{k+1}) = m-1+d(v_{i_{k+1}+1})+d(w_{i_{k+1}})-m-1 = d(v_{i_{k+1}+1})+d(w_{i_{k+1}})-2$ for $k = 0, 1, \ldots, p-1$) and the proof of theorem is complete. QED

Finally, there are of course other problems concerning tree-decompositions, where the porcupine technique can be applied. The well-known conjecture by Gỳarfàs and Lehel (Gỳarfàs& Lehel(1978)) that every list $T_1, T_2, \ldots, T_n$ packs K_n when T_i is a tree with i vertices, $i = 1, 2, \ldots, n$ is an obvious target.

3. The Oberwolfach problem

By the Oberwolfach problem for a graph G I shall mean the problem to determine the set of 2-factors which decompose G. The original variant of this problem as posed by Gerhard Ringel in 1967 (see Guy (1971)) has $G = K_{2n+1}$, but the question makes perfect sense for any $2k$-regular graph, and, what is perhaps more important, can occasionally be answered in something resembling full generality. In fact, we can often hope to say something about the possible lists of 2-factors which decompose the given graph G if, say, G is dense (i.e has many edges) and large. In this case the generic situation is, or rather ought to be, that every proper list of 2-factors packs and although no theorem to this effect is known so far, even when G is complete, there is certainly every reason to believe that this will change eventually.

It has also been suggested as a research problem [Alspach (1981)]to decide which, presumably all, proper list of cycles packs K_{2n+1} and $K_n(2)$, respectively; we may of course hope for general theorems for any large dense eulerian graph G, not only $G = K_{2n+1}$ or $G = K_n(2)$. These look like hard problems to settle in full generality, and they probably are, but some large special cases admit easy solutions.

First of all, let us consider a particular kind of list where every entry occurs an even number of times (such a list will be referred to as *repeated*). Moreover we shall use the term *bilist* for a list where every entry is a bipartite graph. The *repeated bilist-2-factorization problem for a graph* G is of course the problem of determining which proper repeated lists of bipartite 2-regular graphs decompose G. It is clear by a look at figure 2 that $C_n(2)$ admits a B-factorization for any bipartite 2-regular $2n$-order graph B. This can be reformulated as: "Every proper repeated bilist of 2-regular graphs decomposes $C_n(2)$.", and that in turn gives immediately the following lemma from (Häggkvist(1985)):

Lemma 3.1: *Let G be a graph with a hamilton decomposition. Then every proper repeated list of bipartite 2-regular graphs on $2n$ vertices decomposes $G(2)$.*

An easy consequence of this lemma is that both the Oberwolfach problem and the repeated bilist-2-factorization problem can be solved for $K_{4n,4n}$ and $K_{2n+1}(2)$. However, the Oberwolfach problem is still open for $K_{4n+2,4n+2}$, say, as is the problem of determining the set of bipartite 2-factors which decompose $K_{2n}(2)$. See [Alspach&Häggkvist(1985)]for some relevant results.

The ease by which $C_n(2)$ could be decomposed into given bipartite 2-factors suggests that sharper results may not be too hard to obtain for graphs of the form $C_n(2k)$, $k = 2, 3, \ldots$.

Conjecture 3.2: *Every proper list of bipartite 2-factors decomposes $C_n(2k)$ for $k = 2, 3, \ldots$.*

Another glance at figure 2 gives immediately that if the the graph G with m edges has an euler tour $v_1 v_2 \cdots v_m v_1$ (the v_is are vertices in G, not necessarily distinct) such that for every i, $= 1, 2, \ldots, m-p$, the segment $v_i v_{i+1} \cdots v_{i+p}$ induces a path or cycle in G, then every proper repeated bilist of cycles no longer than $2p$ decomposes $G(2)$. A slight variation of this idea was used in (Häggkvist(1985)) to show that every proper repeated list of even cycles of lengths differing from $4n$ packs $K_{2n+1}(2)$, and that similarly every proper repeated list of even cycles of lengths differing from $8n-6$, $8n-4$ or $8n-2$ decomposes $K_{4n,4n}$.

Consequently it would be of interest to investigate the following question.

Problem 3.3: *Let G be an eulerian graph with minimum degree δ. Show that there exists a function $f(\delta)$ such that G admits an eulerian tour $v_1 v_2 \ldots v_m v_1$ such that every segment $v_i v_{i+1} \ldots v_{i+j}$ induces a path or cycle if $j = 1, 2, \ldots, f(\delta)$.*

4. Main results

The purpose of this section is to present some theorems varying the following theme:

For every natural number k there exists a smallest natural number c_k such that every k-regular $2n$-order bipartite graph B decomposes $K_{c_k n, c_k n}$.

It is clear that c_k, if it exists, is a multiple of k. Of course, if we only consider graphs B where $k|n$ we could hope for a B-decomposition of $K_{n,n}$, at least if n is large, and by putting on a wishing cap we can easily formulate more general statements which look very plausible, although perhaps not too easy to prove. Here are two examples.

Conjecture 4.1: *For every natural number k there exists a positive constant ε_k such that every k-regular n-order graph G decomposes every km-regular n-order graph H if $n \leq (1+\varepsilon_k)km$.*

Conjecture 4.2(the bipartite variant): *For every natural number k there exists a positive constant ε_k such that every k-regular $2n$-order bipartite graph B decomposes every $2n$-order km-regular bipartite graph if $n \leq (1+\varepsilon_k)km$.*

Similar conjectures can of course be made for list-decompositions.

I do not currently know of any 3-regular bipartite graph on $6m$ vertices which fails to decompose $K_{3m,3m}$ for $m \geq 3$ and I am inclined to believe that none exists. In fact there is probably only one exception when $m = 2$. Similarily, although there certainly exist small values of r for which certain 3-regular $(3r+1)$-order graphs fail to decompose K_{3r+1} (the Petersen graph does not decompose K_{10} to take a random but perhaps not unexpected example) I strongly suspect that no large (say $r \geq 5$?) r has this property.

The reason I bother to state the bipartite variant separately is that this case seems much easier, in view of the results in the current paper. Here I shall exclusively deal with a variant of the bipartite case where the graph to be decomposed is complete bipartite.

Let us first consider a special class of bipartite graphs which are easy to pack into a complete bipartite graph. First of all recall that a *hypergraph H* consists of a set $V(H)$ of vertices and some family $E(H)$ of subsets (here called hyperedges) of vertices. A hypergraph is k-uniform if every hyperedge has exactly k vertices, k-partite if the vertices can be partitioned into sets $V_1, V_2, \ldots, V_k$ such that every hyperedge has at most one vertex from each V_i, and k-equipartite if in addition the

V_i's are of equal size. (This definition covers our application, but in general various types of loops are allowed). The vertex-edge incidence graph $B = B(V, E)$ of a hypergraph H is the bipartite graph with vertices $V(H) \cup E(H)$ and edges (v, e) if and only if $v \in e$ and $e \in E(H)$. Let $\mathcal{H}(k, m, n)$ denote the family of vertex-edge incidence graphs of k-uniform k-equipartite hypergraphs on kn vertices and containing m hyperedges. Note that every member $B = B(S, T)$ of $\mathcal{H}(k, m, n)$ has a natural k-partition of S as $S_1, S_2, \ldots, S_k$ such that $B[S_i, T]$ consists of a forest of vertex-disjoint stars centered at the vertices in S_i and with altogether m edges. Another way of formulating the same thing is to say that the vertices in S have been coloured using k colours such that every vertex in T is adjacent to all colours (exactly once) and such that every colour is used the same number of times. Note that it it may happen that two vertices in S differ in degree (this never happens for pair of vertices in T).

We now have a simple theorem, which qualifies as an observation. A look at figure 4 says why.

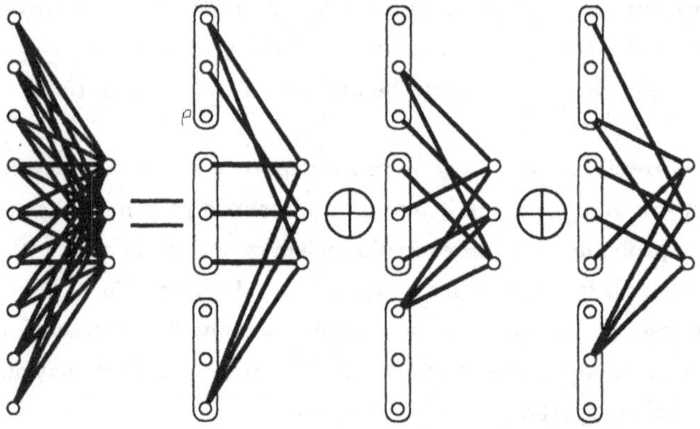

A particular case of theorem 4.3. $B \in \mathcal{H}(3,3,3) \Rightarrow B | K_{9,3}$

Figure 4.

Theorem 4.3: *Every $B = B(S, T) \in \mathcal{H}(k, m, n)$ decomposes $K_{kn,m}$.*

Proof: By assumption $|S| = kn$, $|T| = m$ and S has an equipartition $S_1, S_2, \ldots, S_m$ such that $B[S_i, T]$ consists of a forest of stars centered at S_i. We want to find n labelled copies $B_1, B_2, \ldots, B_n$ of B in such a way that every possible edge st shall occur in exactly one B_p, $p = 1, 2, \ldots, n$. Let T have vertices $t_1, t_2, \ldots, t_m$ and let the vertex corresponding to t_j in B_p be just t_j, for $p = 1, 2, \ldots, n$, $j = 1, 2, \ldots, m$.

It remains to label the vertices in S. The easiest way to proceed is to let the vertices in S_i be $s_1^i, s_2^i, \ldots, s_n^i$ for $i = 1, 2, \ldots, k$, and let the vertex corresponding to s_j^i in B_p be denoted by $s_j^{i,p}$. Put

$$s_j^{i,p} = s_{j+p \pmod n}^i \qquad \text{for every } i, j \text{ and } p.$$

Then, since every vertex in T is joined to exactly one vertex in S_i for $i = 1, 2, \ldots, k$, every possible edge $s_j^i t_q$ must be present in exactly one of the n copies of B. QED

Let us analyse more in detail what happens in the above proof. As we shall see, not only does $K_{kn,m}$ admit one B-decomposition, but indeed many. Let $L(n)$ denote the number of $n \times n$ latin squares, and let $S_1, S_2, \ldots, S_k$ be an equipartition of S, as in the proof of theorem 4.3. We are going to count the number of ways to n-edge-colour (i.e we use as colours the numbers $1, 2, \ldots, n$ and differ between two colourings where the names of the colours are interchanged) the edges of the complete bipartite graph $K = K(S, T)$ such that every colour p induces a copy B_p of B; moreover we require that if the edge $s_j^i t$ with one end in S_i and the other the vertex $t \in T$ is present in B then the corresponding edge in B_p joins some vertex $s_j^{i,p}$ in S_i to t. Such an edge-colouring is said to be a B-decomposition which fixes T and respects the partition of S.

Theorem 4.4: *Let $B = B(S,T) \in \mathcal{H}(k, m, n)$ be the graph in theorem 4.3, but assume furthermore B has no isolated vertices. Then there exist exactly $L(n)^k$ B-decompositions of $K = K(S,T)$ which fix T and respect the partition of S.*

Proof: As in the proof of theorem 4.3 we let the vertices in T be $t_1, t_2, \ldots, t_m$ and the vertices in S_i are still $s_1^i, s_2^i, \ldots, s_n^i$ for $i = 1, 2, \ldots, k$. Let $A^1, A^2, \ldots, A^k$ be a given sequence of $n \times n$ latin squares such that A^i has as symbols the vertices in S_i for $i = 1, 2, \ldots, k$. We shall see that there exists a natural (bijective)correspondence between the partition-respecting B-decompositions of K which fix T and such sequences of latin squares. To see this it suffices to put

$$s_l^{i,p} = A_{l,p}^i. \tag{4.5}$$

where $A_{l,p}^i$ denotes the symbol in cell (l, p) in A^i. In this way, if we let $s_l^{i,j}$ be the vertex in B_j corresponding to the vertex s_l^i in B so that $s_p^{i,j} t_q$ is an edge in B_j if and only if $s_p^i t_q$ is an edge in B, then

i) all vertices $s_l^{i,j}$ are distinct $l = 1,2,\dots,n$ since no symbol occurs twice in column j of A^i

ii) no edge $s_p^i t_q$ in K belongs to more than one B_j since every pair of vertices in S_i have non-overlapping neighbour sets in B

and

iii) every edge is used in some B_j since the degrees tally.

On the other hand, given a partition-respecting B-decomposition of K fixing T, we can obviously use 4.5 to define a latin square A^i on S_i. (In other words, if $s_p^{i,j}t_q$ is an edge in B_j then $A^i_{p,q} = s^i_p$. This is the place where the assumption about the degrees in B is used. Without it we could not define all of A^i. The verificaton that A^i is latin is left to the reader.) QED

The final batch of theorems can be viewed as analogues of Wilson's theorem, but in the realm of decompositions of complete bipartite graphs. The proofs shall make use of standard graph theoretical theorems such as Brooks theorem stating that every graph of maximum degree at most k with no component a K_{k+1} has a proper vertex-k-colouring if $k > 2$.

We shall also make use of the square G^2 of a graph G, i.e the graph obtained by joining every pair of vertices in G of distance at most 2 by an edge.

Theorem 4.5: *Let G be a 3-regular bipartite graph on $2n$ vertices such that no component of G is a Heawood graph. Then $6G|K_{6n,6n}$.*

Proof: Note that since no component of the bipartite 3-regular graph $G = G[S,T]$ is isomorphic with a Heawood graph we have that $G^2[S]$ is a graph with maximum degree less than 7 and without K_7 as a component. Consequently, by Brooks theorem, $G^2[S]$ admits a proper 6-vertex-colouring. Let the colour classes of one such colouring Λ be $S_1, S_2, \dots, S_6$ so that in particular $G[S_i, T]$ apart from isolated vertices in T consists of a set of 3-stars centered at S_i for $i = 1,2,\dots,6$. For every permutation π of $1,2,\dots,6$ we let $G(\pi)$ denote the copy of G for which the colours $1,2,\dots,6$ have been changed to $\pi(1),\pi(2),\dots,\pi(6)$ respectively. Now consider the cyclic permutation $\sigma = (6,1,2,\dots,5)$ and the reflection $\tau = (1,6,5,4,3,2)$ and the 12 permutations $\sigma^i, \sigma^i\tau$ for $i = 1,2,\dots,6$.

Put $G_i = G(\sigma^i)$ and $\tilde{G}_i = G(\sigma^i\tau)$ for $i = 1,2,\dots,6$ and take a look at figure 5. This particular figure is meant to show the twelve graphs in question with their respective colourings. The position of the graphs is indicated in the accompanying text. Each graph is drawn with the T-vertices in the central oval. The spokes indicate possible edges, and (this is important) the numbers in the diagram indicate where the vertices in each S_i are located in the particular copy of G. Thus in particular the total set of vertices in colour 2 in G_2, G_4 and G_6 consists of the copies of S_6, S_4 and S_2 in that order.

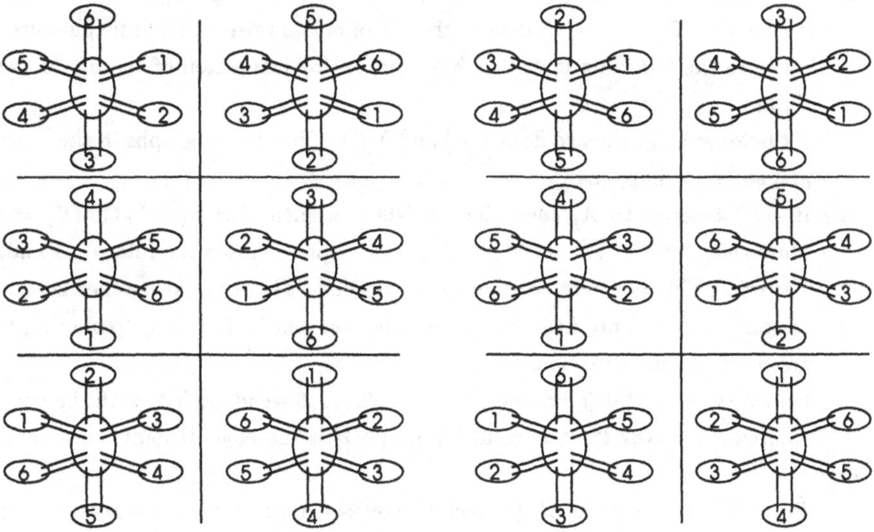

$G_1 \quad G_2 \quad \tilde{G}_1 \quad \tilde{G}_2$
$G_3 \quad G_4 \quad \tilde{G}_3 \quad \tilde{G}_4$
$G_5 \quad G_6 \quad \tilde{G}_5 \quad \tilde{G}_6$
Figure 5.

The vertices in T are partitioned into equivalence classes by the following rule. $t \in T$ belongs to T_i if the colours of the neighbours in G is one of the 3-sets in A_i for $i = 1, 2, 3$ where:

$$A_1 = \left\{ \begin{matrix} \{1,2,3\}, & \{2,3,4\}, & \{3,4,5\} \\ \{4,5,6\}, & \{5,6,1\}, & \{6,1,2\} \end{matrix} \right\}$$

$$A_2 = \left\{ \begin{matrix} \{1,3,4\}, & \{2,4,5\}, & \{3,5,6\}, & \{4,6,1\}, & \{5,1,2\}, & \{6,2,3\} \\ \{2,5,6\}, & \{3,6,1\}, & \{4,1,2\}, & \{5,2,3\}, & \{6,3,4\} & \{1,4,5\} \end{matrix} \right\}$$

$$A_3 = \left\{ \begin{matrix} \{1,3,5\} \\ \{2,4,6\} \end{matrix} \right\}$$

We shall now take the graphs $G_1 \cup G_2 \cup \cdots \cup G_6$ and $\tilde{G}_1 \cup \tilde{G}_2, \cdots \cup \tilde{G}_6$ and identify T-vertices in such a way that the resulting graph belongs to $\mathcal{H}(6, 6n, 2n)$. After that we shall identify pair of vertices in S so that the resulting graph belongs to $\mathcal{H}(6, 6n, n)$ and admits a $6G$-decomposition. Finally, an application of theorem 4.3 finishes the proof. The T-identification is done now.

For each vertex t in T we denote by t^i and $\tilde{t}^i$ the corresponding vertex in G_i and $\tilde{G}_i$ respectively. Moreover for each vertex x of the in total $12n$ copies of T-vertices in the copies of G we let $\lambda(x)$ denote the set of colours used on the neighbours of x. Now note the following facts which are proved by inspection of the figures 6 to 10. Fix an i, $i = 1, 2, \ldots, 6$

i) If t belongs to T_j then so does $\lambda(\tilde{t}^i)$ and $\lambda(t^i)$. (Rotate the graphs in the figures and see what happens.)

ii) If $\lambda(t^i)$ belongs to A_j then there exists a k such that $\lambda(\tilde{t}^{i+k}) \cap \lambda(t^i) = \emptyset$. Moreover, $\lambda(\tilde{t}^{i+k+1}) \cap \lambda(t^{i+1}) = \emptyset$ (superscripts counted modulo 6). Finally, k is odd. This is where the figures are needed. There are altogether 20 cases to consider, and only five are shown. The rest can be found by exploiting the symmetry of the situation.

For every vertex t in T and every i, $i = 1, 2, \ldots, 6$ we identify t^i with the vertex $\tilde{t}^{i+k}$ to create a vertex $\hat{t}^i$. The resulting graph $\hat{B}$ is the edge-disjoint union of the graphs

$\hat{B}_1$ and $\hat{B}_2$ where $\hat{B}_1 \simeq 6G$ (in fact $\hat{B}_1$ comes from $G_1 \cup G_2 \cup \cdots \cup G_6$) and the same goes for $\hat{B}_2$ (which is $\tilde{G}_1 \cup \tilde{G}_2 \cup \cdots \cup \tilde{G}_6$).

Moreover $\hat{B} = \hat{B}(\hat{S}, \hat{T})$ has $12n$ $\hat{S}$-vertices and $6n$ $\hat{T}$-vertices. The $\hat{S}$ vertices have all degree 3 as in G whereas every vertex in $\hat{T}$ has degree 6. Finally, and this is the most important thing, the vertices in $\hat{S}$ are coloured in 6 colours such that every colour is used exactly $2n$ times (because all the vertices $\hat{T}$ are adjacent to all six colours exactly once) and moreover, if $\hat{S}_i$ are the $2n$ vertices of colour i in $\hat{S}$ then n of these come from $\hat{B}_1$, i.e the graph $G_1 \cup G_2 \cup \cdots \cup G_6$ whereas the other n come from the graph $\hat{B}_2 = \tilde{G}_1 \cup \tilde{G}_2 \cdots \cup \tilde{G}_6$. Now match in each $\hat{S}_i$ every vertex from $\hat{B}_1$ to the vertices in $\hat{B}_2$ and identify the paired vertices (the identification can thus be made in $n!$ ways in each $\hat{S}_i$). The resulting graph B^* belongs to $\mathcal{H}(6, 6n, n)$ and admits a $6G$-decomposition. The theorem follows, since by theorem 4.3 $K_{6n,6n}$ admits a B^*-decomposition. QED

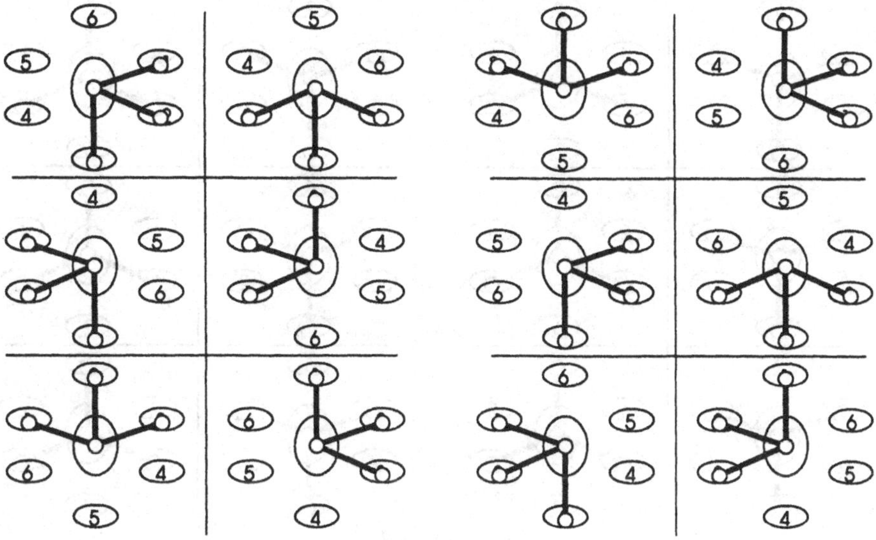

Figure 6.

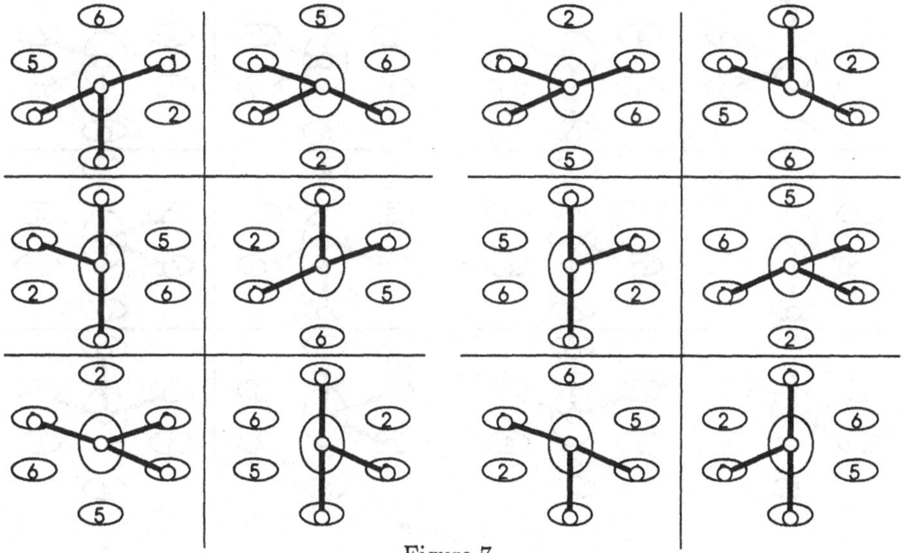

Figure 7.

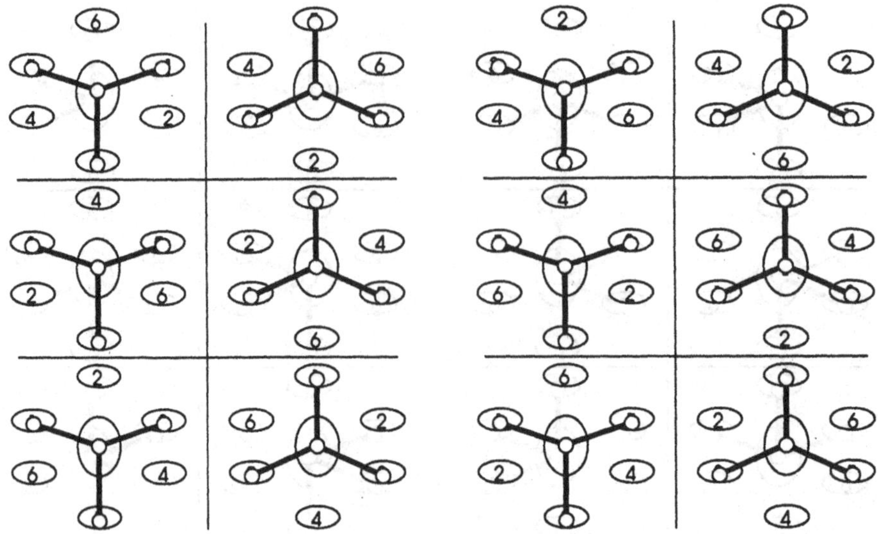

Figure 8.

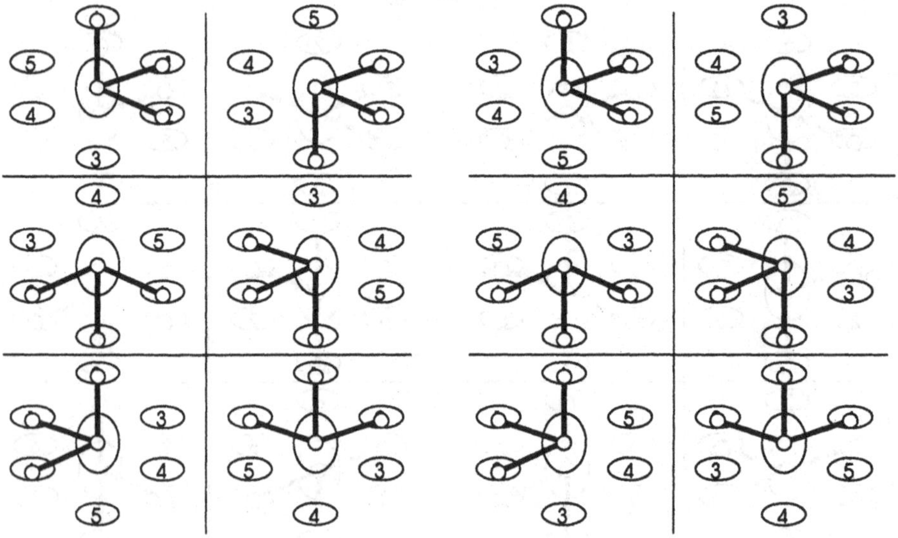

Figure 9.

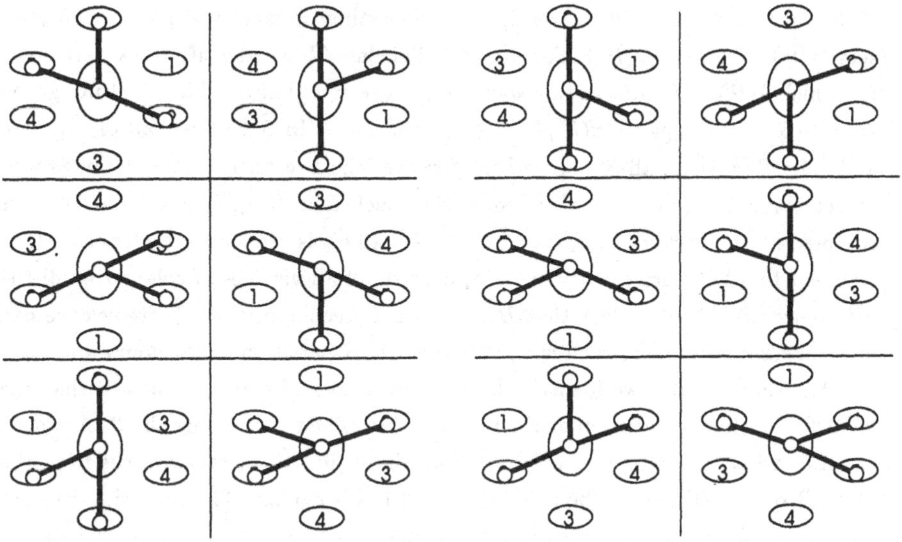

Figure 10.

The next theorem is entirely similar in its proof except that the identifications at the end must be given in a more general setting, and the reader is well advised to treat the theorem as an exercise.

Theorem 4.6: *Every k-regular bipartite graph on $2n$ vertices decomposes $K_{k^2!n,k^2!n}$*

Proof: The proof is initially much the same as the proof of theorem 4.5, but the result is not as sharp. Let G be a bipartite k-regular graph on $2n$ vertices, n of which belong to the part S. Note that the maximum vertex degree in $G^2[S]$ is no larger than $k(k-1) \leq k^2 - 1$ so that $G^2[S]$ admits a proper vertex-k^2-colouring with colour classes $S_1, S_2, \ldots, S_{k^2}$. With notation as in the proof of theorem 4.5 we consider the $k^2!$ graphs $G(\pi)$ where π is a permutation of $1, 2, \ldots, k^2$. List the graphs $G(\pi)$ as $G_1, G_2, \ldots, G_{k^2!}$ arbitrarily. We keep the notation $\lambda(t^i)$ for the set of colours on the neighbours of the T-vertex t^i in the graph G_i where t^i is the vertex corresponding to the vertex t in G. For any fixed T-vertex t in G we have that the hypergraph with vertices the colours $1, 2, \ldots, k^2$ and edges the family of sets $(\lambda(t^i) : i = 1, 2, \ldots, k^2!)$ is a complete k-uniform hypergraph on k^2 vertices, where each hyperedge has the same multiplicity. It follows from a celebrated theorem of Baranyai's (Baranyai(1973)) (but this is really much too heavy artillery to use) that the sets $\lambda(t^i)$, $i = 1, 2, \ldots, k^2!$ can be partitioned into parallel classes, say

$T_1, T_2 \dots, T_q$, each parallel class T_p, say, consisting of exactly k pairwise disjoint sets $\lambda(t^{i_1}), \lambda(t^{i_2}), \dots, \lambda(t^{i_k})$. For each parallel class T_p we identify the k vertices t^j for which $\lambda(t^j) \in T_p$ and in this way we get a new vertex $\tilde{t}^p$. The resulting graph $\hat{B} = \hat{B}(\hat{S}, \hat{T})$ belongs to $\mathcal{H}(k^2, \frac{k^2!}{k}n, \frac{k^2!}{k^2}n)$ and has $k^2!n$ $\hat{S}$-vertices, all of degree k and $\frac{k^2!}{k}n$ $\hat{T}$-vertices, all of degree k^2. Moreover the $\hat{S}$-vertices are partitioned into k^2 sets $\hat{S}_1, \hat{S}_2, \dots, \hat{S}_{k^2}$, say, all of equal size, such that $\hat{B}[\hat{S}_i, \hat{T}]$ is a forest of stars centered at the vertices in $\hat{S}_i$, $i = 1, 2, \dots, k^2$. (This is seen since each vertex $\tilde{t}^p$ is adjacent to all colours exactly once. $\hat{S}_i$ consists of all vertices of colour i in all the $k^2!$ copies of G). It is clear that $\hat{B}$ admits a G-decomposition. Moreover we can now take k copies of $\hat{B}$, each with its canonical $\hat{S}$ partition, (i.e colours are kept) and for each colour i we identify the vertices in lots of k in such a way that the identified vertices come from distinct copies of $\hat{B}$. The resulting graph B^* obviously has a G-decomposition (recall that we have never identified two vertices from the same $\hat{B}$ or initially, from the same G_i.) and is k^2-regular. The colouring has not changed as far as the $\hat{T}$-vertices are concerned, so therefore the graph belongs to $\mathcal{H}(k^2, k^2!n, \frac{k^2!}{k^2}n)$. Again this implies that B^* packs $K_{k^2!n, k^2!n}$ by theorem 4.3 and we have proved 4.6.

Theorem 4.7: *Let $G = G(S,T)$ be a bipartite 3-regular graph on $2n$ vertices, n even, and assume that every vertex in S belongs to a 4-cycle. Then $3G|K_{3n,3n}$. Moreover $G|K_{n,3n}$ if $6|n$.*

Proof: The proof is very similar to the proof of theorem 4.5 above. However, instead of giving a 6-vertex colouring of $G^2[S]$ we note that the maximum degree in $(G^2[S])$ is 5 or less, whence there exists a 6-vertex-colouring of $G^2[S]$ such that each colour class has order $\lfloor \frac{1}{6}n \rfloor$ or $\lfloor \frac{1}{6}n \rfloor + 1$ by the Szemeredi-Hajnal theorem [see e.g th.5.1′ in Bollobás (1978)].

This means in particular that $G^2[S]$ has a vertex colouring where the total number of vertices of colours 1,3 or 5 equals the total number of vertices of colours 2,4 or 6 (i.e $\frac{1}{2}n$, –recall that n is even). Using the graphs G_1, G_3 and G_5 together with $\tilde{G}_2, \tilde{G}_4$ and $\tilde{G}_6$ and the identification $t^i, \tilde{t}^{i+k}$ as before gives us $\hat{B}$ as in the proof of theorem 4.5. We note this time that $\hat{B} = \hat{B}_1 \oplus \hat{B}_2$ where $\hat{B}_1$ is a copy of $G_1 \cup G_3 \cup G_5$ while $\hat{B}_2 = \tilde{G}_2 \cup \tilde{G}_4 \cup \tilde{G}_6$. It remains to identify the vertices in $\hat{S}_i$ for $i = 1, 2, \dots, 6$.

Note that $\hat{S}_i$ consists of $\frac{1}{2}n$ vertices from $G_1 \cup G_3 \cup G_5[S]$ and $\frac{1}{2}n$ vertices from $\tilde{G}_2 \cup \tilde{G}_4 \cup \tilde{G}_6[S]$. If we match the the first kind of vertices to the second to produce $\frac{1}{2}n$ new vertices $\hat{s}^i$ then the resulting graph has a $3G$-decomposition and belongs to

$\mathcal{H}(6, 3n, 3n)$ and thus decomposes $K_{3n,3n}$ by theorem 4.5. Thus $3G|K_{3n,3n}$ if n is even. Moreover, if $6|n$ then each $\hat{S}_i$ consists of $\frac{1}{6}n$ vertices from each of the original 6 colour sets and we just identify the vertices in lots of 6, one from each of the original colour sets. In this way we avoid the risk of identifying two vertices from the same $\tilde{G}_i$, $i = 2, 4, 6$ or from the same G_i, $i = 1, 3, 5$. Consequently the resulting graph B^* admits a G-decomposition. Furthermore, since every T-vertex in the B^* has degree 6 and still is adjacent to all of the 6 colours it must be the case that B^* belongs to $\mathcal{H}(6, 3n, \frac{1}{2}n)$ and therefore packs $K_{n,3n}$. QED

Theorem 4.8: *Every bipartite 3-regular graph on $2n$ vertices with no Heawood-graph as component decomposes $K_{3n,3n}$ if n is even.*

Proof: We again revert to the proof of theorem 4.5, this time word for word, except that we use the graphs G_1, G_3, and G_5 together with the graphs $\tilde{G}_2, \tilde{G}_4$ and $\tilde{G}_6$. The identifications giving the vertex $\hat{t}^i$ are performed as before, and we end up with a graph in $\mathcal{H}(6, 3n, n)$. However, we can no longer be certain that $\frac{1}{2}n$ of the vertices of colour i in $\hat{B}$, (i.e the vertices in $\hat{S}_i$), stem from $G_1 \cup G_3 \cup G_5$. All we know is that exactly one copy of every vertex in S appears in $\hat{S}_i$ for $i = 1, 2, \ldots, 6$. The problem we face is to identify pairs of vertices in $\hat{S}_i$ without identifying two vertices from the same copy of G. Fortunately this is the same as finding a perfect matching in the graph L_i on $\hat{S}_i$ whose edges are given by $xy \in E(L_i)$ if and only if x and y do not belong to the same copy of G. However, this just means that the graph L_i is complete 6-partite, and since no copy of G has more than $\frac{1}{2}n$ vertices in $\hat{S}_i$ L_i has a perfect matching. (In fact no copy of G can have more than $\frac{1}{3}n$ vertices in $\hat{S}_i$ since G is 3-regular and bipartite with $2n$ vertices so that any set of more than $\frac{1}{3}n$ vertices in S must contain some pair of vertices of distance 2, and no such pair of vertices can be found in $\hat{S}_i$.) The theorem follows. QED

Theorem 4.9: *Every bipartite k-regular graph on $2n$ vertices decomposes $K_{2k^2!^2n^2+1}$*
Theorem 4.10: *Every 3-regular bipartite graph on $2n$ vertices with no Heawood graph as component decomposes K_{18n^2+1} if n is even.*

Proofs: Just note that $K_{m,m}$ has a graceful labelling for every m (easy exercise), and this immediately implies that $K_{m,m}$ decomposes K_{2m^2+1} as well. An application of theorem 4.6 and 4.8 gives immediately 4.9 and 4.10.QED

References

Brian Alspach: *Problem no.3*, Discrete Mathematics 36(1981), 333-334.

Brian Alspach and Roland Häggkvist: *Some observations on the Oberwolfach Problem*, J.Graph Theory 9(1985), 177–187.

Zsolt Baranyai: *On the factorization of the complete uniform hypergraph*, Coll. Math. Soc. János Bolyai 10. (Infinite and finite sets, Keszthely (Hungary),1973, ed. A.Hajnal, R. Rado and Vera T. Sós), North Holland (1975), 91–109 (ISBN 0 7204 2814 9).

Bela Bollobás: *Extremal graph theory*, Academic Press 1978 (ISBN 0 12 111750 2).

A.G. Chetwynd and A.J.W. Hilton: *Regular graphs of high degree are 1-factorizable*, Proc. London Math. Soc. (3), 50 (1985), 193—206.

J.Dénes and A.D.Keedwell: *Latin squares and their applications*, The English Universities Press 1974 (ISBN 0 340 12489 X).

Torbjörn Gustavsson: *Finding a triangle decomposition of a large graph with high minimum degree*, Report from the department of mathematics, University of Stockholm, Sweden, 1987- No.35.

Richard K.Guy: *Unsolved Combinatorial Problems*, Combinatorial Mathematics and its Applications, ed D.Welsh, Academic Press 1971 (ISBN 0 12 743350 3).

A. Gỳarfàs and J.Lehel: *Packing trees of different order into* K_n, Colloquia Mathematica Societatis János Bolyai 18 (1978), 463–469.

Roland Häggkvist: *A lemma on cycle decompositions*, Annals of Discrete Mathematics 27(1985)(Cycles in graphs, ed. B.R.Alspach and C.D.Godsil), 227–232 (ISBN 0 444 87803 3).

Roland Häggkvist:*Restricted edge-colourings of bipartite graphs*, manuscript (1982).

Marshall Hall,Jr: *Combinatorial Constructions*, MAA Studies of Mathematics vol.17 (Studies in Combinatorics ed. Gian-Carlo Rota) (1978), pp. 218 - 253 (ISBN 0 88385 117 2).

C. Huang, A. Kotzig and A. Rosa: *Further results on tree labellings* , Utilitas Mathematica Vol 21C (1982), pp. 31-48.

Denes König: *Über Graphen und ihre Anwendung auf Determinantentheorie und Mengenlehre*, Mathematische Annalen 77 (1916), 453–465.

C.W.H. Lam, L.H.Thiel and S. Swiers: *A computer search for a projective plane of order* 10, Algebraic, Extremal and Metric Combinatorics 1986, ed. Deza,

Frankl and Rosenberg, London Mathematical Society Lecture Note Series 131, Cambridge University Press 1988, pp 155—165.

C.St.J.A. Nash-Williams: untitled problem in the book Combinatorial Theory and its applications III, Coll. Math. Soc. János Bolyai 4, ed. P.Erdös, A. Rényi and Vera T. Sós, North Holland 1970, (ISBN 7204 2037 7), 1179-1182.

Gerhard Ringel: *Problem no.*25, Theory of graphs and its applications, (Proceedings of the Symposium held in Smolenice in June 1963, ed. Miroslav Fiedler), Publishing house of the Czechoslovak Academy of Sciences, Prague 1964.

Anne Penfold Street and Deborah J. Street: *Combinatorics of experimental design*, Oxford Science Publications 1987, (ISBN 0 19 853256 3 and ISBN 0 19 853255 5 (pbk)).

G.Tarry: *Le problème des* 36 *officiers*, C.R. Assoc. France Av. Sci. 29(1900), part 2, 170-203.

Richard M. Wilson: *Decompositions of complete graphs into subgraphs isomorphic to a given graph*, Proc. 5th British Combinatorial Conference 1975, pp. 647–659.

Department of Mathematics, University of Umeå, S-90187 Umeå, Sweden

ON THE METHOD OF BOUNDED DIFFERENCES

Colin McDiarmid

§1 Introduction

In the beginning Maurey (1979) used an inequality for bounded martingale difference sequences to prove an isoperimetric inequality for the symmetric group S_n; this isoperimetric inequality was then used in investigating normed spaces,for related work see Milman & Schechtman (1986). Then Shamir & Spencer (1987) and Rhee & Talagrand (1987) introduced this 'bounded differences' method to a wide public of researchers in combinatorics and the mathematics of operational research and computer science, with dramatic impact. The purpose of this paper is to discuss the method and some of these applications. The underlying martingale result is due to Hoeffding (1963), Azuma (1967) and has often been referred to as 'Azuma's inequality'.

In this introductory section we first present a simple version of a bounded differences inequality which does not require us to introduce any concepts like martingales. We then sketch a plan of the rest of the paper. First recall the Chernoff bound on the tails of the binomial distribution.

(1.1) Lemma (Chernoff (1952)): Let $X_1,\ldots,X_n$ be independent random variables, with $P[X_k=1] = p$ and $P[X_k=0] = 1-p$ for each k. Then for any $t > 0$

$$P\left[\,|\Sigma X_k - np| \geq t\right] \leq 2\exp\left[-2t^2/n\right].$$

The sum here is of course over k from 1 to n. This will always be the case when we write an unadorned sum Σ or product Π.

The above inequality has proved its usefulness many times in combinatorics and elsewhere (see for example Erdős & Spencer (1974)). There are other times when one would like to use it but the corresponding object of interest will not quite separate into independent parts. What can we do? Perhaps we can use the following 'independent bounded differences inequality'.

(1.2) Lemma: Let $X_1,\dots,X_n$ be independent random variables, with X_k taking values in a set A_k for each k. Suppose that the (measurable) function $f: \Pi A_k \to \mathbb{R}$ satisfies

(1.3) $$|f(\underline{x}) - f(\underline{x}')| \le c_k$$

whenever the vectors $\underline{x}$ and $\underline{x}'$ differ only in the kth co-ordinate. Let Y be the random variable $f\left[X_1,\dots,X_n\right]$. Then for any $t > 0$,

$$P(|Y - E(Y)| \ge t) \le 2\exp\left[-2t^2/\Sigma c_k^2\right].$$

If we take each $A_k = \{0,1\}$ and $f(\underline{x}) = \Sigma x_k$ we may obtain lemma (1.1). Often we shall take A_k as a set of edges in a graph, as for example in lemmas (3.1) to (3.3) below.

The plan of this paper is as follows. In the next section we show part of the recent impact of these inequalities by sketching 'before' and 'after' pictures of our knowledge about colouring random graphs. Then we sketch proofs of the 'after' theorems using lemmas derived from lemma (1.2) above.

In the next section, section 4, we introduce martingales and present the basic 'Azuma's inequality' of Hoeffding (1963), Azuma (1967). (This does not quite yield lemma (1.2) above.)

In section 5 we present the 'Hoeffding' family of inequalities for sums of independent bounded random variables, which extend the Chernoff bound. These results generalise to inequalities involving martingales, and in section 6 we give these generalisations (and prove lemma (1.2)). These results in section 6 extend

Azuma's inequality.

In sections 7 and 8 we briefly sketch several further applications of these inequalities concerning isoperimetric inequalities in graphs and certain problems in operational research and computer science, and we conclude in section 9.

§2 Colouring random graphs – before and after

In this section we show the dramatic effect of the bounded differences method on our knowledge about the stability number and the chromatic number of a random graph. (We cannot talk about colourings without talking about stable sets.)

For a splendid introduction to the theory of random graphs see Bollobás (1985). Recall that the random graph $G_{n,p}$ has vertex set $\{1,...,n\}$ and the $\binom{n}{2}$ possible edges occur independently with probability $p = p(n)$. For simplicity we shall focus mainly on the case p constant, with $0<p<1$. Let $q = 1-p$ and $b = 1/q$. Recall also that a set of vertices in a graph G is stable if no two are adjacent; and the stability number $\alpha(G)$ is the largest size of such a set. A colouring of G is a partition of the vertices into stable sets; and the chromatic number $\chi(G)$ is the least number of blocks in such a partition.

The stability number $\alpha\left[G_{n,p}\right]$ is concentrated on only a few values (see Matula (1970), (1972), (1976) Grimmett & McDiarmid (1975), Bollobás & Erdős (1976)). Indeed (when p is constant) there is a function $r = r(n,p)$ (see section 3(c) below) such that

(2.1) $$P\left[r-1 \le \alpha\left[G_{n,p}\right] \le r\right] \to 1 \text{ as } n \to \infty.$$

This result was proved by 'classical' first and second moment methods (see Bollobás (1985)). Our knowledge about the stability number of sparse random graphs was patchy (again see Bollobás (1985)).

Now for colouring. Grimmett and McDiarmid (1975) (see also Bollobás & Erdős (1976) showed that (when p is constant) the chromatic number $\chi\left[G_{n,p}\right]$

satisfies

(2.2) $$\frac{n}{2\log_b n}(1+o(1)) \le \chi[G_{n,p}] \le \frac{n}{\log_b n}(1+o(1))$$

for almost all graph $G_{n,p}$ (that is, with probability $\to 1$ as $n \to \infty$). Also they conjectured that the lower bound was the 'correct' value. For results about sparse random graphs and related results see McDiarmid (1984). There was no significant improvement on the 1975 result (2.2) until Matula (1987) introduced a factor 2/3 into the upper bound, by using an elegant 'expose–and–merge' method for generating a random graph.

That was the 'before' picture, now for the 'after'.

The first breakthrough was when Shamir and Spencer in 1987 showed that the chromatic number $\chi[G_{n,p}]$ is highly concentrated. Let us say that the graph function X is <u>concentrated in width</u> $s = s(n,p)$ if there exists a function $u = u(n,p)$ so that

$$P[u \le X[G_{n,p}] \le u+s] \to 1 \text{ as } n \to \infty.$$

Also let us use $\omega(n)$ to denote a function which tends to ∞ arbitrarily slowly as $n \to \infty$.

(2.3) <u>Theorem</u> (Shamir & Spencer (1987)): For any probability function $p = p(n)$, $\chi[G_{n,p}]$ is concentrated in width $n^{1/2}\omega(n)$.

(2.4) <u>Theorem</u> (Shamir & Spencer (1987)): Let $p = n^{-\alpha}$, where $0 < \alpha < 1$.
i) For $0<\alpha<1/2$, $\chi[G_{n,p}]$ is concentrated in width $s = n^{(1/2)-\alpha}\omega(n)$.
ii) For $1/2<\alpha<1$, $\chi[G_{n,p}]$ is concentrated in width $s = \lceil(2\alpha+1)/(2\alpha-1)\rceil$.

Shamir and Spencer comment that it is both a strength and a weakness of the method that u is not given explicitly. Luczak (1988a) reports considerable strengthenings of (2.3), (2.4) again based on the bounded differences method.

A year later Bollobás (1988a) gave the outstanding application of the bounded differences method. He showed roughly that for the function r in (2.1) above,

(2.5) $$P\left[\alpha\left[G_{n,p}\right] < r-2\right] \text{ is tiny compared to } 2^{-n}.$$

Given (2.1) and a suitable form of (2.5) it will be easy to handle colourings, and obtain

(2.6) <u>Theorem</u>: For almost all graphs $G_{n,p}$

$$\chi\left[G_{n,p}\right] = (1+o(1))n/2\log_b n.$$

This of course establishes the conjecture in Grimmett & McDiarmid (1975). (Korsunov (1980) put forward a proof for $p = 1/2$ but this was incomplete and hard to follow.) Bollobás actually proved a more precise theorem; and reworking his arguments carefully (for this paper!) shows the following even more precise result.

(2.7) <u>Theorem</u>: For almost all graphs $G_{n,p}$

$$\chi\left[G_{n,p}\right] = n/\left[2\log_b n - 2\log_b\log_b n + 0(1)\right].$$

Similar methods may be used to handle the chromatic number of (i) moderately sparse random graphs $G_{n,p}$ (Bollobás (1988a), Luczak (1988b)); (ii) random graphs $G_{n,k-out}$ where k is about cn (Bollobás (1988c)); and (iii) random uniform hypergraphs (Bollobás (1988c), Shamir (1988b), (1988c)). (To form the random graph $G_{n,k-out}$ each of the vertices 1,..,n independently directs edges to k other vertices at random, then directions are ignored and any multiple edges combined.)

Now let us turn back to stable sets. Frieze (1989) uses the bounded differences method to pin down the stability number of sparse random graphs. A corollary of his theorem is the following.

(2.8) <u>Theorem</u>: Let $p = p(n)$ be given, write $d = d(n) = np$, and suppose that

$d(n) \to \infty$ but $d(n) = o(n)$ as $n \to \infty$. Then for almost all graphs $G_{n,p}$

$$\alpha\left[G_{n,p}\right] = \frac{2n}{d}(\log d - \log\log d - \log 2 + 1 + o(1)).$$

§3 Colouring random graphs – proofs

(a) General lemmas

The following useful results are all special cases of lemma (1.2). They are implicit in Shamir & Spencer (1987) (see also Bollobás (1988c)) though we have improved the bounds here.

(3.1) Lemma: Let $\left[A_1,\ldots,A_m\right]$ be a partition of the edge–set of the complete graph K_n into m blocks; and suppose that the graph function f satisfies $|f(G) - f(G')| \leq 1$ whenever the symmetric difference $E(G)\Delta E(G')$ is contained in a single block A_k. Then the random variable $Y = f\left[G_{n,p}\right]$ satisfies

$$P(|Y - E(Y)| \geq t) \leq 2\exp\left[-2t^2/m\right] \qquad \text{for } t > 0.$$

The above result lemma (3.1) follows directly from lemma (1.2) with each $c_k = 1$. The next two results follow directly from lemma (3.1). For the former let A_k be the set of edges $\{j,k\}$ where $j < k$; and for the latter let the blocks A_k be singletons.

(3.2) Lemma: Suppose that the graph function f satisfies $|f(G)-f(G')| \leq 1$ whenever G' can be obtained from G by changing edges incident with a single vertex. Then the corresponding random variable $Y = f\left[G_{n,p}\right]$ satisfies

$$P(|Y - E(Y)| \geq t) \leq 2\exp\left[-2t^2/n\right] \quad \text{for } t > 0.$$

(3.3) Lemma: Suppose that the graph function f satisfies $|f(G)-f(G')| \leq 1$ whenever G and G' differ in only one edge. Then the corresponding random variable $Y = f\left[G_{n,p}\right]$ satisfies

$$P(|Y - E(Y)| \geq t) \leq 2 \exp\left[-2t^2/\binom{n}{2}\right] \qquad \text{for } t > 0.$$

(b) Concentration of $\chi[G_{n,p}]$

By lemma (3.2) above we deduce immediately that if Y is the random variable $\chi[G_{n,p}]$ then for any $t > 0$

(3.4) $$P(|Y - E(Y)| \geq t) \leq 2\exp[-2t^2/n].$$

This proves theorem (2.3) (and is a sharpening of (28) in [SS]).

Clearly the same inequality will hold if Y is $\Psi[G_{n,p}]$ where $\Psi(G)$ is the achromatic number (see McDiarmid (1982) for the asymptotic behaviour of $\Psi[G_{n,p}]$); or Y is $i[G_{n,p}]$ where $i(G)$ is the interval number of G (Scheinerman (1989)); or in many other examples.

Now let us consider a second proof of theorem (2.3) which will also prove theorem (2.4). This proof method is due to Alan Frieze, see also Luczak (1988a). We thus avoid the complications of the proof in Shamir & Spencer (1987) of theorem (2.4) (though the inequality theorem 7 in that paper may be of some interest).

Let $\omega(n) \to \infty$ arbitrarily slowly, and let $u = u(n)$ be the least integer such that

(3.5) $$P[\chi[G_{n,p}] \leq u] \geq 1/\omega(n).$$

For a graph with n vertices, let f(G) be the least size of a subset W of vertices of G such that $\chi(G\backslash W) \leq u(n)$. Let Y be the random variable $f[G_{n,p}]$. Then by lemma (3.2)

(3.6) $$P(|Y - E(Y)| \geq t) \leq 2\exp[-2t^2/n] \qquad \text{for } t > 0.$$

But by (3.5), $P(Y = 0) \geq 1/\omega(n)$, so we must have $E(Y) \leq n^{1/2}\omega(n)$ for large n; and now by (3.6) we have

(3.7) $$P[Y < 2n^{1/2}\omega(n)] \to 1 \quad \text{as} \quad n \to \infty.$$

Theorem (2.3) follows easily from (3.5) and (3.7). However, we can also now prove theorem (2.4) (which is a slight sharpening of the original result). We need only show the following lemma.

(3.8) Lemma: With probability tending to 1, every set of at most $n^{1/2}\omega(n)$ vertices in $G_{n,p}$ may be s–coloured, where s is given in theorem (2.4).

We may prove lemma (3.8) by showing that the expected number of sets as above with each degree at least s–1 tends to 0 as $n \to \infty$, as in Shamir & Spencer (1987).

(c) Stable sets in $G_{n,p}$

Here we present a suitable explicit form of (2.5), namely lemma (3.11) below, following the ideas in Bollobás (1988a).

Given an integer r, $1 \le r \le n$, let E(n,r) be the expected number of stable sets of size r in $G_{n,p}$. Thus $E(n,r) = \binom{n}{r} q^{\binom{r}{2}}$. Next for real r, $1 \le r \le n$, let

$$\hat{E}(n,r) = (2\pi)^{-1/2} n^{n+(1/2)} (n-r)^{-(n-r+(1/2))} r^{-(r+(1/2))} q^{r(r-1)/2}.$$

By Stirling's formula, $E(n,r) = (1+o(1))\hat{E}(n,r)$ if $r=r(n)$ is an integer, $r(n) \to \infty$ but $r(n)=0(\log n)$. Now define

(3.9) $$r = r(n) = 2\log_b n - 2\log_b \log_b n + 2\log_b \left[\frac{e}{2}\right].$$

Then straightforward calculations show that for any $t = t(n) = 0(1)$,

(3.10) $$\hat{E}(n, r+t) = n^{1-t+0(\log\log n/\log n)}.$$

It follows that

$$P\left[\alpha\left[G_{n,p}\right] < r + 1 + 1/\log\log n\right] \to 1 \text{ as } n \to \infty.$$

We shall show that

(3.11) Lemma: $P\left[\alpha\left[G_{n,p}\right] < \lfloor r - 1/2 - 1/\log\log n \rfloor\right] = o\left[\exp\left[-n^{1+1/\log\log n}\right]\right].$

Proof: Set $t(n) = -(1/2) - 1/\log\log n$. Then by (3.10)

$$\hat{E}(n, r+t) = n^{3/2+(1+o(1))/\log\log n}.$$

Let $s = s(n) = \lfloor r(n) + t(n) \rfloor$. Further calculations now show that

(3.12) $$\hat{E}(n', s(n)) = n^{3/2+(1+o(1))/\log\log n}$$

for some integer $n' = n'(n)$ with $n\left[q^{1/2} + o(1)\right] \le n' \le n$.

We are going to use lemma (3.3), so we shall need an appropriate function f. This is the clever trick. For a graph G of order n we define f(G) to be the maximum number of sets in a collection $S_1, S_2, \dots$ of stable sets of size s(n) with $|S_i \cap S_j| \le 1$ for $i \ne j$. Of course $|f(G) - f(G')| \le 1$ if G and G′ differ only in one edge, so we can indeed apply lemma (3.3). Let Y be the random variable $f\left[G_{n,p}\right]$. Then

(3.13) $$P(|Y - E(Y)| \ge t) \le 2\exp\left[-4t^2/n^2\right] \qquad \text{for } t > 0.$$

But $P\left[\alpha\left[G_{n,p}\right] < s\right] = P(Y = 0)$, so

(3.14) $$P\left[\alpha\left[G_{n,p}\right] < s\right] \le 2\exp\left[-4E(Y)^2/n^2\right].$$

To use this inequality we want a good lower bound on E(Y). For graphs G on {1,...,n} define f′(G) to be the number of stable sets $S \subseteq \{1,\dots,n'\}$ of size s(n) which are such that for any other such set S′ we have $|S \cap S'| \le 1$. Of course $E(Y) \ge E\left[f'\left[G_{n,p}\right]\right]$. Further calculations show that $E\left[f'\left[G_{n,p}\right]\right] = (1+o(1))E(n', s(n))$, and hence lemma (3.11) follows from (3.12) and (3.14). □

(d) Colouring $G_{n,p}$

Let us see here how lemma (3.11) on large stable sets allows us to pin down $\chi\left[G_{n,p}\right]$. The approach was sketched out in Bollobás and Erdős (1976) (section 5(i)), but it was not until Bollobás used the bounded differences method to prove a result like lemma (3.11) that it could be made to work. We shall sketch a proof of the following result, which of course yields theorem (2.7). A detailed proof is given in McDiarmid (1989).

(3.15) <u>Theorem</u>: Let $0 < p = 1-q < 1$ be fixed, let $b = 1/q$, and let

$$r = r(n) = 2\log_b n - 2\log_b\log_b n + 2\log_b\left[\frac{e}{2}\right],$$

as before. Then for almost all graphs $G_{n,p}$

$$n/\{r+o(1)\} \le \chi\left[G_{n,p}\right] \le n/\left\{r - \tfrac{1}{2} - \frac{1}{1-q^{1/2}} + o(1)\right\}.$$

The lower bound follows from an easy first moment calculation which need not concern us here. The upper bound follows from the two lemmas below, one concerning random graphs and one deterministic.

Let us say that a graph G has property Q_n if it has n nodes and for all subsets W of at least $n/\log^3 n$ nodes we have $\alpha(G[W]) \ge s(|W|)$. Here G[W] denotes the subgraph of G induced by W, and the function s is defined in the proof of lemma (3.11).

(3.16) <u>Lemma</u>: Almost all graphs $G_{n,p}$ have property Q_n.

(3.17) <u>Lemma</u>: Consider (deterministic) graphs G_n with property Q_n for n = 1,2,... Then

$$\chi\left[G_n\right] \le n/\left\{r(n) - \tfrac{1}{2} - \frac{1}{1-q^{1/2}} + o(1)\right\}.$$

The deterministic lemma (3.17) follows by considering repeatedly picking out large stable sets until less than $n/\log^3 n$ nodes remain, and then colouring the remaining nodes all with different colours.

<u>Proof of lemma (3.16):</u>

By lemma (3.11), for n sufficiently large

$$\begin{aligned} &P\{G_{n,p} \text{ does not have } Q_n\} \\ \le\ & 2^n \exp\left[-\left[n/\log^3 n\right]^{1+(1/2\log\log n)}\right] \\ \le\ & 2^n e^{-n} \to 0 \text{ as } n \to \infty. \end{aligned}$$

□

(e) Stability number of sparse random graphs

The upper bound in theorem (2.8) from Frieze (1989) is straightforward (see Bollobás (1985) lemma XI.21). Let us sketch the lines of the lower bound proof.

For a suitable function $n' = n'(n)$ let $\mathscr{P}_n$ be a partition of $\{1,...,n\}$ into n' almost equal–sized sets. (In Frieze (1989) n' is about $n(\log d)^2/d$.) In a graph G with n vertices, call a set S of vertices $\mathscr{P}$–stable if it is stable and $|S \cap B| \leq 1$ for each block B of $\mathscr{P}_n$. Let $\beta(G)$ be the largest size of a $\mathscr{P}$–stable set. Of course $\beta(G) \leq \alpha(G)$. (Frieze attributes the idea of considering $\mathscr{P}$–stable sets rather than just stable sets to Luczak.)

Let Y_n be the random variable $\beta\left[G_{n,p}\right]$. Lemma (3.1) shows that for any $t > 0$

(3.18) $$P\left[|Y_n - E\left[Y_n\right]| \geq t\right] \leq 2\exp\left[-2t^2/n'\right].$$

But now let

$$k = k(n) = \frac{2n}{d}(\log d - \log\log d - \log 2 + 1 - \epsilon).$$

Second moment calculations on the number of $\mathscr{P}$–stable sets of size k show that $P\left[Y_n \geq k\right]$ does not tend to zero very quickly. It then follows from (3.18) that $E\left[Y_n\right] \geq k - (\epsilon n/d)$, and the lower bound in theorem (2.8) follows by using (3.18) again.

§4 Martingales

Let $(\Omega, \mathscr{F}, P)$ be a probability triple. Given a random variable X and a sub–σ–field $\mathscr{G}$ of $\mathscr{F}$, we shall use the notation $E(X|\mathscr{G})$ to denote the expectation of X conditional on $\mathscr{G}$. In most of our applications Ω will be a finite set. There is then a partition of Ω such that the σ–field $\mathscr{F}$ is the collection of sets which are unions of blocks of the partition. A random variable is a real–valued function X defined on Ω such that X is constant on the blocks. Also $E(X|\mathscr{G})$ is simply the function f defined on Ω which is constant on the blocks, the constant on each block being the average value of X on the block.

A nested sequence $\mathscr{F}_0 = \{\phi, \Omega\} \subseteq \mathscr{F}_1 \subseteq \mathscr{F}_2 \subseteq \ldots \subseteq \mathscr{F}$ of sub–σ–fields of $\mathscr{F}$ is called a <u>filter</u>. In the finite case this corresponds to a sequence of increasingly refined partitions of Ω, starting with the trivial partition with one block Ω. Given a filter, a sequence $X_0, X_1, X_2, \ldots$ of integrable random variables is called a <u>martingale</u> if $E[X_{n+1} | \mathcal{F}_n] = X_n$ for each $n \geq 0$. Also, given a filter, a sequence $Y_1, Y_2, \ldots$ of integrable random variables is called a <u>martingale difference sequence</u> (mds) if for each $n \geq 1$, Y_n is $\mathcal{F}_n$–measurable and $E[Y_n | \mathcal{F}_{n-1}] = 0$.

From a martingale $X_0, X_1, X_2, \ldots$ we obtain a martingale difference sequence by setting $Y_k = X_k - X_{k-1}$; and conversely from X_0 and a martingale difference sequence $Y_1, Y_2, \ldots$ we obtain a martingale $X_0, X_1, X_2, \ldots$ by setting $X_k = X_0 + \sum_{i=1}^{k} Y_i$. Thus we may focus on either sequence.

We shall be interested here only in finite filters $\mathcal{F} = \{\phi, \Omega\} \subseteq \mathcal{F}_1 \subseteq \ldots \subseteq \mathcal{F}_n \subseteq \mathcal{F}$. All corresponding martingales $X_0, X_1, \ldots, X_n$ are then obtained by Doob's martingale process: let X be an integrable random variable and define $X_k = E[X | \mathcal{F}_k]$ for $k = 0, \ldots, n$. Thus $X_0 = E(X)$ and $X_n = X$ if X is $\mathcal{F}_n$–measurable. Then if $Y_1, \ldots, Y_n$ is the corresponding martingale difference sequence we of course have $X - EX = \Sigma Y_k$.

The basic inequality for a bounded martingale difference sequence is the following lemma of Hoeffding (1963), Azuma (1967) (see also Freedman (1975)), which has often been referred to as 'Azuma's inequality'.

(4.1) <u>Lemma</u>: Let $Y_1, \ldots, Y_n$ be a martingale difference sequence with $|Y_k| \leq c_k$ for each k, for suitable constants c_k. Then for any $t > 0$,

$$P\left[|\Sigma Y_k| \geq t\right] \leq 2\exp\left[-t^2/2\Sigma c_k^2\right].$$

Suppose as in (1.1) that $X_1, \ldots, X_n$ are independent, with $P[X_k = 1] = p$ and $P[X_k = 0] = 1-p$. Set $Y_k = X_k - p$ and $c_k = \max(p, 1-p)$, and apply lemma (4.1) to obtain the Chernoff bound (1.1), except that the bound is weakened slightly (if $p \neq 1/2$). We may also deduce a similar weakened version of lemma (1.2). All our applications here will in fact be based on less symmetrical forms of lemma (4.1) and will thus avoid a gratuitous factor 1/4 in the exponent in previously presented bounds.

We shall prove stronger results in section 6 below (see corollary (6.9)) but it is convenient to prove lemma (4.1) here. First we prove one preliminary lemma.

(4.2) <u>Lemma</u>: Let the random variable Y satisfy $E(Y) = 0$ and $|Y| \le 1$. Then for any $h > 0$

$$E\left[e^{hY}\right] \le e^{h^2/2}.$$

<u>Proof</u>: As e^{hx} is a convex function of x,

$$e^{hx} \le \frac{1-x}{2} e^{-h} + \frac{1+x}{2} e^{h} \quad \text{for } -1 \le x \le 1,$$

and so

$$E\left[e^{hY}\right] \le \tfrac{1}{2}\left[e^{-h}+e^{h}\right] \le e^{h^2/2},$$

by considering series expansions. □

<u>Proof of lemma (4.1):</u>

Let $S_k = \sum_{i=1}^{k} Y_i$. By the Markov or Bernstein inequality, for any $h > 0$

$$\text{Prob}\left[S_n \ge t\right] \le e^{-ht} E\left[e^{hS_n}\right].$$

But

$$\begin{aligned} E\left[e^{hS_n}\right] &= E\left[e^{hS_{n-1}} e^{hY_n}\right] \\ &= E\left[e^{hS_{n-1}} E\left[e^{hY_n} \mid \mathcal{F}_{n-1}\right]\right] \\ &\le E\left[e^{hS_{n-1}}\right] \exp\left[\tfrac{1}{2}\left[hc_n\right]^2\right] \quad \text{by lemma (4.2)} \\ &\le \exp\left[\tfrac{1}{2}h^2 \Sigma c_k^2\right] \quad \text{on iterating.} \end{aligned}$$

Now set $h = t/\Sigma c_k^2$ to obtain

$$\text{Prob}\left[S_n \ge t\right] \le \exp\left[-t^2/2\Sigma c_k^2\right].$$

By symmetry the absolute value gives at most a factor two.

□

For a similar simple proof of a weaker version of the inequality see Milman & Schechtman (1986). Above we used our bounds on $|Y_k|$ to obtain good bounds on $E\left[e^{hY_k} | \mathcal{F}_{k-1}\right]$. Other assumptions which yield bounds on this latter quantity will yield related inequalities – see for example Johnson et al (1985), Bollobás (1988c) theorem 7.

§5 Inequalities for bounded independent summands

In this section we present the 'Hoeffding' family of inequalities for sums of independent bounded random variables. In the next section we extend these to inequalities involving martingales.

(a) Results

We shall take the following inequality as our starting point.

(5.1) Theorem (Hoeffding (1963)): Let the random variables $X_1,...,X_n$ be independent, with $0 \leq X_k \leq 1$ for each k. Let $\overline{X} = \frac{1}{n}\Sigma X_k$, $p = E[\overline{X}]$, and $q = 1-p$. Then for $0 \leq t < q$,

$$P(\overline{X} - p \geq t) \leq \left[\left[\frac{p}{p+t}\right]^{p+t}\left[\frac{q}{q-t}\right]^{q-t}\right]^n$$

A special case of interest is when each random variable X_k is 1 with probability p and 0 with probability q. The theorem then reduces to a bound on a tail of the binomial distribution due to Chernoff (1952) though the methods involved date back to Bernstein (see Chvátal (1979)). This bound is good for large deviations (see Chernoff (1952), Bahadur & Ranga Rao (1960), Bahadur (1971)) though inequalities closer to the normal approximation of DeMoivre – Laplace are naturally better for small deviations (see Feller (1968), Bollobás (1985)). (For related inequalities concerned with variations of the p_i subject to $\Sigma p_i = p$ see Hoeffding (1956), Gleser (1975). The bound applies also to the hypergeometric distribution – see Hoeffding (1963), Chvátal (1979).) It is straightforward to obtain from theorem (5.1) weaker but more useful bounds.

(5.2) Corollary: As in theorem (5.1), let the random variables $X_1,\dots,X_n$ be independent, with $0 \le X_k \le 1$ for each k, let $\bar{X} = \frac{1}{n}\Sigma X_k$ and let $p = E[\bar{X}]$.

(a) For $t > 0$,

(5.3) $$P(\bar{X} - p \ge t) \le \exp\left[-2nt^2\right],$$

(5.4) $$P(\bar{X} - p \le -t) \le \exp\left[-2nt^2\right].$$

(b) For $0 < \epsilon < 1$

(5.5) $$P(\bar{X} - p \ge \epsilon p) \le \exp\left[-\tfrac{1}{3}\epsilon^2 np\right],$$

(5.6) $$P(\bar{X} - p \le -\epsilon p) \le \exp\left[-\tfrac{1}{2}\epsilon^2 np\right].$$

The inequalities in (a) are perhaps the basic workhorses, but often p is small in applications and then the inequalities in (b) may be better. Part (a) is due to Hoeffding (1963) who also discusses relationships between theorem (5.1), the inequalities in (a) and other similar inequalities. Part (b) appears in Angluin & Valiant (1979) (in the binomial case). In Karp (1979) the inequality (5.3) is attributed to Angluin. For similar results (in the binomial case) based on Stirling's approximation see Bollobás (1985) Chapter 1, corollary 4 and theorem 7(i).

Hoeffding also gives a powerful extension of corollary (5.2) (a) to the case when the ranges of the summands need not be the same.

(5.7) Theorem: Let the random variables $X_1,\dots,X_n$ be independent, with $a_k \le X_k \le b_k$ for each k, for suitable constants a_k, b_k. As before let $\bar{X} = \frac{1}{n}\Sigma X_k$ and $p = E[\bar{X}]$. Then for $t > 0$

$$P(\bar{X} - p \ge t) \le \exp\left[-2n^2t^2/\Sigma\left[b_k - a_k\right]^2\right],$$

$$P(\bar{X} - p \le -t) \le \exp\left[-2n^2t^2/\Sigma\left[b_k - a_k\right]^2\right].$$

(b) Proofs

Let us prove the above results here, as the proofs will be useful also for the extensions of these results in the next section (and proofs of the much used inequalities (5.5) and (5.6) do not seem to be easily available in the literature).

Proof of theorem (5.1):

Let $m = (p + t)n$. For $s \geq 1$,

$$\begin{aligned} P\left[\Sigma X_k \geq m\right] &\leq E\left[s^{\Sigma X_k - m}\right] \\ &= s^{-m} \Pi\, E\left[s^{X_k}\right] \\ &= s^{-m} \Pi \left[q_k + p_k s\right] \\ &\leq s^{-m}(q + ps)^n, \end{aligned}$$

since geometric means are at most arithmetic means. Now set $s = \frac{(p+t)q}{p(q-t)}$ to obtain the desired inequality. □

Proof of corollary (5.2):

(a) Let $f(t) = (p+t)\ln \frac{p}{p+t} + (q-t)\ln \frac{q}{q-t}$ for $-p < t < q$. Then

$$f'(t) = \ln \frac{p(q-t)}{(p+t)q},$$

and

$$f''(t) = -\frac{1}{(p+t)(q-t)} \leq -4$$

since

$$(p+t)(1-(p+t)) \leq \tfrac{1}{4}.$$

But $f(0) = f'(0) = 0$, and so it follows from Taylor's theorem that for $0 \leq t < q$

$$\begin{aligned} f(t) &= \left[t^2/2\right] f''(s) \qquad \text{for some } s,\ 0 \leq s \leq t \\ &\leq -2t^2. \end{aligned}$$

The inequality (5.3) now follows from theorem (5.1), and by considering $1-X_k$ we obtain (5.4).

(b) Now let $g(x) = f(xp)$, for $0 \leq x \leq 1$, $xp < q$. Then $g'(x) = pf'(xp)$ and

$$g''(x) = p^2 f''(xp) = -\frac{p}{(1+x)(q-xp)} \leq -\frac{p}{1+x}.$$

Thus

$$g'(x) \le -p\ln(1+x)$$

$$\le -(2p/3)x \qquad \text{since } 0 \le x \le 1,$$

and so

$$g(x) \le -(p/3)x^2.$$

Together with theorem (5.1) this yields (5.5).

Finally let $h(x) = g(-x)$ for $0 \le x < 1$. Then $h'(x) = -g'(-x)$ and $h''(x) = g''(-x)$. Thus $h(0) = h'(0) = 0$ and

$$h''(x) = -\frac{p}{(1-x)(q+xp)} \le -p,$$

so $h(x) \le -px^2/2$ and (5.6) follows from theorem (5.1).

□

It remains here only to prove theorem (5.7). But once we have proved lemma (5.8) below which extends lemma (4.2), then theorem (5.7) will follow by a proof like that of lemma (4.1).

(5.8) Lemma: Let X be a random variable with $E(X) = 0$, $a \le X \le b$. Then for $h > 0$

$$E\left[e^{hX}\right] \le \exp\left[\tfrac{1}{8}h^2(b-a)^2\right].$$

Proof Arguing as for lemma (4.2)

$$e^{hx} \le \frac{x-a}{b-a}e^{hb} + \frac{b-x}{b-a}e^{ha} \qquad \text{for } a \le x \le b,$$

and so

$$E\left[e^{hX}\right] \le \frac{b}{b-a}e^{ha} - \frac{a}{b-a}e^{hb} \qquad = e^{f(\hat{h})} \tag{5.9}$$

where $\hat{h} = h(b-a)$ and $f(x) = -px + \log\left[1-p+pe^x\right]$.

Now we may argue as in the proof of corollary (5.2). We have

$$f'(x) = -p + \frac{p}{p+(1-p)e^{-x}}$$

and

$$f''(x) = \frac{p(1-p)e^{-x}}{\left[p+(1-p)e^{-x}\right]^2} \le \frac{1}{4}.$$

Also $f(0) = f'(0) = 0$, so

$$f(\hat{h}) \leq \tfrac{1}{8}\hat{h}^2 = \tfrac{1}{8}h^2(b-a)^2. \qquad \square$$

§6 Inequalities for bounded martingale difference sequences

(a) Results

Here we present the 'Hoeffding' family of inequalities for bounded martingale difference sequences. These extend the basic inequality (4.1) ('Azuma's inequality') and the inequalities of section 5 for the independent case. The main thrust of these results was pointed out in Hoeffding (1963) and we largely follow his treatment. We shall also give an extension of lemma (1.2). The use of any of these inequalities in combinatorics is what we call the 'bounded differences' method.

(6.1) Theorem: Let $Y_1,\ldots,Y_n$ be a martingale difference sequence with $-a_k \leq Y_k \leq 1-a_k$ for each k, for suitable constants a_k. Let $a = \frac{1}{n}\Sigma a_k$ and $\bar{a} = 1 - a$. Then for any $t > 0$

$$P\left[\Sigma Y_k \geq nt\right] \leq \left[\left[\frac{a}{a+t}\right]^{a+t}\left[\frac{\bar{a}}{\bar{a}-t}\right]^{\bar{a}-t}\right]^n.$$

To obtain theorem (5.1) from theorem (6.1) set $a_k = E\left[X_k\right]$ and $Y_k = X_k - a_k$. We shall prove theorem (6.1) shortly. As before we want weaker but more useful bounds. The same proof as for the corollary (5.2) to theorem (5.1) will yield

(6.2) Corollary: As above, let $Y_1,\ldots,Y_n$ be a martingale difference sequence with $-a_k \leq Y_k \leq 1-a_k$ for each k, for suitable constants a_k; and let $a = \frac{1}{n}\Sigma a_k$.

(a) For any $t > 0$,

(6.3) $$P\left[\Sigma Y_k \geq t\right] \leq \exp\left[-2t^2/n\right],$$

(6.4) $$P\left[\Sigma Y_k \le -t\right] \le \exp\left[-2t^2/n\right].$$

(b) For any $0 \le \epsilon \le 1$,

(6.5) $$P\left[\Sigma Y_k \ge \epsilon a\right] \le \exp\left[-\tfrac{1}{3}\epsilon^2 a/n\right],$$

(6.6) $$P\left[\Sigma Y_k \le -\epsilon a\right] \le \exp\left[-\tfrac{1}{2}\epsilon^2 a/n\right].$$

To obtain corollary (5.2) from corollary (6.2) set $a_k = E\left[X_k\right]$ and $Y_k = X_k - a_k$ as before.

The next result extends inequalities (6.3) and (6.4) above and theorem (5.7) for the independent case, and Azuma's inequality lemma (4.1). It seems to be the most useful inequality for the bounded differences method. We shall state it in terms of martingales, with a corollary in terms of martingales difference sequences and a corollary extending lemma (1.2).

(6.7) <u>Theorem</u>: Let $(\phi, \Omega) = \mathcal{F}_0 \subseteq \mathcal{F}_1 \subseteq \ldots \subseteq \mathcal{F}_n$ be a filter. Let the integrable random variable X be $\mathcal{F}_n$–measurable, and let $X_0, X_1, \ldots, X_n$ be the martingale obtained by setting $X_k = E\left[X | \mathcal{F}_k\right]$. Suppose that for each k=1,...,n there is a constant c_k and an $\mathcal{F}_{k-1}$–measurable function a_k such that

(6.8) $$a_k \le X_k \le a_k + c_k.$$

Then for any $t > 0$,

$$P(X - EX \ge t) \le \exp\left[-2t^2/\Sigma c_k^2\right],$$
$$P(X - EX \le -t) \le \exp\left[-2t^2/\Sigma c_k^2\right].$$

The next result is essentially the special case with each function a_k constant. It extends the basic inequality (4.1) (Azuma's inequality).

(6.9) <u>Corollary</u>: Let $Y_1, \ldots, Y_n$ be a martingale difference sequence with $a_k \le Y_k \le b_k$ for each k, for suitable constants a_k, b_k. Then for any $t > 0$

$$P\left[\Sigma Y_k \ge t\right] \le \exp\left[-2t^2/\Sigma\left[b_k - a_k\right]^2\right],$$
$$P\left[\Sigma Y_k \le -t\right] \le \exp\left[-2t^2/\Sigma\left[b_k - a_k\right]^2\right].$$

The last result here will also follow easily from theorem (6.7). It is an extension of lemma (1.2) in which the condition (1.3) is replaced by a weaker but less attractive condition.

(6.10) Corollary: Let $Z_1,\dots,Z_n$ be random variables with Z_k taking values in a set A_k, and let $\underline{Z}$ denote the random vector $\left(Z_1,\dots,Z_n\right)$. Let $f: \Pi A_k \to \mathbb{R}$ be an appropriately measurable function. Suppose that there are constants $c_1,\dots,c_n$ so that

(6.11)
$$|E\left(f(\underline{Z})|\left(Z_1,\dots,Z_{k-1}\right)=\left(z_1,\dots,z_{k-1}\right), Z_k=z_k\right)$$
$$-E\left(f(\underline{Z})|\left(Z_1,\dots,Z_{k-1}\right)=\left(z_1,\dots,z_{k-1}\right), Z_k=z_k'\right)| \le c_k$$

for each $k = 1,\dots,n$ and $z_i \epsilon A_i$ $(i = 1,\dots,k-1)$ and z_k, $z_k' \epsilon A_k$. Then for any $t > 0$,
$$P(|f(\underline{Z}) - Ef(\underline{Z})| \ge t) \le 2\exp\left(-2t^2/\Sigma c_k^2\right).$$

(b) Proofs

Proof of theorem (6.1)

We combine the proofs of theorems (4.1) and (5.1). Let $S_k = \sum_{i=1}^{k} Y_i$.

For any $h > 0$,
$$\begin{aligned}
P\left(S_n \ge nt\right) &\le e^{-hnt}E\left(\exp hS_n\right)\\
&= e^{-hnt}E\left(\exp\left(hS_{n-1}\right)E\left(\exp hY_n|\mathcal{F}_{n-1}\right)\right)\\
&\le e^{-hnt}E\left(\exp\left(hS_{n-1}\right)\right)\left[\left(1-a_n\right)e^{-ha_n}+a_ne^{h\left(1-a_n\right)}\right] && \text{by (6.9)}\\
&\le e^{-hnt}\prod_{k=1}^{n}\left[\left(1-a_k\right)e^{-ha_k}+a_ke^{h\left(1-a_k\right)}\right] && \text{on iterating}\\
&= e^{-hnt}e^{-h\Sigma a_k}\prod_{k=1}^{n}\left(1-a_k+a_ke^h\right)\\
&\le e^{-hnt}e^{-hna}\left(1-a+ae^h\right)^n.
\end{aligned}$$

Now set $e^h = \dfrac{(a+t)\bar{a}}{a(\bar{a}-t)}$ to obtain the desired inequality.

□

Proof of theorem (6.7)

Let $Y_k = X_k - X_{k-1}$ and $S_k = \sum_{i=1}^{k} Y_i = X_k - X_0$. We argue initially as in the proof of theorem (6.1). For any $h > 0$,

$$\begin{aligned} P(X - EX \geq t) &= P\left[S_n \geq t\right] \\ &\leq e^{-ht} E\left[\exp h\, S_n\right] \\ &= e^{-ht} E\left[\exp h\, S_{n-1}\, E\left[\exp hY_n \mid \mathcal{F}_{n-1}\right]\right] \\ &\leq e^{-ht} E\left[\exp h\, S_{n-1}\right] \exp\left[\tfrac{1}{8} h^2 c_n^2\right] \quad \text{by lemma (5.8)} \\ &\leq e^{-ht} \exp\left[\tfrac{1}{8} h^2 \Sigma c_k^2\right] \text{ on iterating.} \end{aligned}$$

Now set $h = 4t/\Sigma c_k^2$ to obtain the former inequality in theorem (6.7). To deduce the latter replace X by $-X$. □

Proof of corollary (6.10):

Let $\mathcal{F}_k$ be the σ-field generated by $Z_1, \ldots, Z_k$. Let $X_k = E\left[f(Z) \mid \mathcal{F}_k\right]$ and let

$$a_k = \text{ess inf}\left[X_k \mid \mathcal{F}_{k-1}\right],\ b_k = \text{ess sup}\left[X_k \mid \mathcal{F}_{k-1}\right].$$

Then the condition (6.11) says that $b_k - a_k \leq c_k$. So corollary (6.10) now follows from theorem (6.7). □

(c) Inequalities for maxima

All the inequalities we have met here are based on the Bernstein inequality

$$P(X \geq t) \leq e^{-ht} E\left[e^{hX}\right] \quad \text{for } h > 0.$$

Thus they can all be strengthened in the following way, as noted in Hoeffding (1963). See also Steiger (1967), (1969), (1970).

Let $Y_1, \ldots, Y_n$ be a martingale difference sequence, let $h > 0$, let $S_k = Y_1 + \ldots + Y_k$ and let $T_k = \exp hS_k$. As long as the T_k are integrable, $T_1, \ldots, T_n$ forms a submartingale and so we can use Doob's maximal inequality (see for example Chung (1974) pages 320 and 330) to deduce that for any $t > 0$,

$$P\left[\max_k S_k \geq t\right] = P\left[\max_k T_k \geq e^{ht}\right]$$
$$\leq e^{-ht}E\left[T_n\right] = e^{-ht}E\left[e^{hS_n}\right].$$

It follows that in all our inequalities in this section we may replace ΣY_k by $\max S_k$, and similarly for independent summands. We do not have applications here for these extensions.

We have already discussed some applications of the bounded differences method to random graphs. See Bollobás (1988b), (1988c) for further applications in this area in particular concerning the first cycle problem and the probability of containing a given small subgraph (and see section 8(d), (e) below). In the last two sections before our concluding remarks we shall discuss two further areas of applications.

§7 Isoperimetric inequalities for graphs

Consider a finite metric space (V,d). We shall be interested in particular in the case when V is the vertex set of a graph G and d measures distance in the graph.

For $A \subseteq V$ and $t>0$ the t–neighbourhood A_t of A is the set $\{v \in V: d(v,A) \leq t\}$. Here $d(v,A)$ is of course $\min\{d(v,x): x \in A\}$. An 'isoperimetric inequality' means a lower bound on $|A_t|$ depending on $|A|$ and t. For an introduction to discrete isoperimetric inequalities see Bollobás (1986).

(a) General results

Following Schechtman (1982), Milman & Schechtman (1986) we make a fussy but useful definition. In the finite metric space (V,d) a partition sequence $\left[\left(\mathcal{P}_k, c_k\right): k=0,\ldots,n\right]$ consists of a sequence $\mathcal{P}_0, \mathcal{P}_1, \ldots, \mathcal{P}_n$ of increasingly refined partitions of V, starting with the trivial partition $\mathcal{P}_0$ (with a single block V) and ending with the discrete partition $\mathcal{P}_n$ (into singleton blocks), and a sequence

$c_0, c_1, ..., c_n$ of numbers with the following property: for each k=1,...,n, whenever $A,B \in \mathcal{P}_k$ and $A,B \subseteq C \in \mathcal{P}_{k-1}$ for some C then there is a bijection $\varphi: A \to B$ with $d(x,\varphi(x)) \leq c_k$ for all $x \in A$.

If the space has such a partition sequence we shall say that it has <u>partition size</u> at most Σc_k^2. Observe that here Σc_k is always at least the diameter $\max\{d(x,y): x,y \in V\}$. Usually we shall have each $c_k = 1$, and indeed we do so in each of the example below.

(7.1) <u>Example</u>: The n–cube Q_n has vertex set $V = \{0,1\}^n$ and two vectors are adjacent if they differ in just one coordinate. Then Q_n has diameter and partition size equal to n. Indeed we may take $\mathcal{P}_k$ as the partition into equivalence classes where two vectors are equivalent if they agree in the first k digits. The appropriate function φ just switches the kth digit.

(7.2) <u>Example</u>: The 'permutation graph' S_n has vertices the permutations of {1,...,n} and two vertices g and h are adjacent if $g^{-1}h$ is a transposition. Then S_n has diameter and partition size equal to n–1. Indeed we may think of a permutation g as a sequence (g(1),...,g(n)), and take $\mathcal{P}_k$ as above for k=0,...,n–1. If $g \in A$, $h \in B$ then the appropriate function φ acts on $x \in A$ by swapping the numbers x(k) = g(k) and h(k).

(7.3) <u>Example</u>: We may generalise the last example as follows (see Schechtman (1982)). The k–tuple graph $S_{n,k}$ has vertices the k–tuples of distinct members of {1,...,n}, and two vertices g and h are adjacent if and only if either they differ in exactly one co–ordinate or they differ in exactly two co–ordinates, say j and k,and g(j) = h(k), g(k) = h(j). Thus distinct vertices g and h are adjacent if and only if they may be extended to permutations which are adjacent in the permutation graph S_n. The graph $S_{n,k}$ has diameter and partition size equal to k if $k \leq n-1$.

The basic result here follows easily from theorem (6.7) or corollary (6.10). It is similar to lemma (1.2), which we took as our workhorse earlier.

(7.4) Theorem: Suppose that the finite metric space (V,d) has a partition sequence $\left[\left[\mathcal{P}_k, c_k\right]: k=1,\ldots,n\right]$. Let the function f on V satisfy $|f(x) - f(y)| \le d(x,y)$ for all $x,y \in V$. Let X be uniformly distributed over V. Then for any $t > 0$,

$$P(f(X) - Ef(X) \ge t) \le \exp\left[-2t^2/\Sigma c_k^2\right],$$
$$P(f(X) - Ef(X) \le -t) \le \exp\left[-2t^2/\Sigma c_k^2\right].$$

(7.5) Corollary: Suppose that the finite metric space (V,d) has a partition sequence $\left[\left[\mathcal{P}_k, c_k\right]: k=1,\ldots,n\right]$. Let $A \subseteq V$ with $|A|/|V| = \alpha$, $0<\alpha<1$. Then for any $t \ge t_0 = \left[\frac{1}{2}\left[\log\frac{1}{\alpha}\right]\Sigma c_k^2\right]^{1/2}$,

$$|A_t|/|V| \ge 1 - \exp\left[-2\left[t-t_0\right]^2/\Sigma c_k^2\right].$$

Thus for any $t > 0$ and any $\gamma > 0$,

(7.6) $$|A_t|/|V| \ge 1 - \left[\frac{1}{\alpha}\right]^{\gamma^2} \exp\left[-2\left[\frac{\gamma}{1+\gamma}\right]^2 t^2/\Sigma c_k^2\right].$$

The above results follow the lines of the extension in Schechtman (1982) of the argument in Maurey (1979) for example (7.2) above – see also Milman & Schechtman (1986), Bollobás (1988c). For applications to the analysis of biased random sources see Shamir (1988a).

Proof of corollary (7.5):

Let $t_1 = Ef(X)$. Then by theorem (7.4)

$$\alpha = P(f(X) = 0) = P\left[f(X) \le t_1 - t_1\right] \le \exp\left[-2t_1^2/\Sigma c_k^2\right],$$

and it follows that $t_1 \le t_0$. Now by theorem (7.4) again, for $t \ge t_0$,

$$\begin{aligned} 1 - |A_t|/|V| &= P(f(X) > t) \\ &\le P\left[f(X) > t_1 + \left[t-t_0\right]\right] \\ &\le \exp\left[-2\left[t-t_0\right]^2/\Sigma c_k^2\right]. \end{aligned}$$

To prove the inequality (7.6) consider separately the cases $t \lesssim (1+\gamma)t_0$ and $t \gtrsim (1+\gamma)t_0$. □

Corollary (7.5) yields remarkable concentration results for the examples (7.1)–(7.3) above. Let us state these results for the n–cube Q_n and the permutation graph S_n.

(7.7) Proposition: Let A be a subset of the 2^n vertices of the n–cube Q_n, with $|A|/2^n = \alpha$, $0<\alpha<1$. Then for $t \geq t_0 = \left[\frac{1}{2}\left[\log\frac{1}{\alpha}\right]n\right]^{1/2}$,

$$|A_t|/2^n \geq 1 - \exp\left[-2\left[t-t_0\right]^2/n\right].$$

Thus for any $t > 0$ and $\gamma > 0$

$$|A_t|/2^n \geq 1 - \left[\frac{1}{\alpha}\right]^{\gamma^2} \exp\left[-2\left[\frac{\gamma}{1+\gamma}\right]^2 t^2/n\right].$$

(7.8) Proposition: Let A be a subset of the n! vertices of the permutation graph S_n, with $|A|/n! = \alpha$, $0 < \alpha < 1$. Then for any $t \geq t_0 = \left[\frac{1}{2}\left[\log\frac{1}{\alpha}\right](n-1)\right]^{1/2}$

$$|A_t|/n! \geq 1 - \exp\left[-2\left[t-t_0\right]^2/(n-1)\right].$$

Thus for any $t > 0$ and $\gamma > 0$

$$|A_t|/n! \geq 1 - \left[\frac{1}{\alpha}\right]^{\gamma^2} \exp\left[-2\left[\frac{\gamma}{1+\gamma}\right]^2 t^2/n\right].$$

Proposition (7.8) is a tightening of the original result of Maurey (1979), though it is still not clear how tight it is – see Bollobás (1987). In order to set these last results in a more general framework let us introduce some definitions. See Milman & Schechtman (1986) for more background. For a graph G with vertex set V and diameter D, and for any $0<\epsilon<1$ let

$$\alpha(G,\epsilon) = \min\{1 - |A_{\epsilon D}|/|V| : A \subseteq V, |A|/|V| \geq 1/2\}.$$

A sequence $G_1, G_2, \ldots$ of graphs is a Lévy family if $\alpha\left[G_n, \epsilon\right] \to 0$ as $n \to \infty$, for every ϵ. It is a concentrated Lévy family if there are $c_1, c_2 > 0$ such that

$\alpha[G_n,\epsilon] \le c_1\exp[-c_2\epsilon n^{1/2}]$ for all n and ϵ. It is a normal Lévy family if there are $c_1,c_2 > 0$ such that $\alpha[G_n,\epsilon] \le c_1\exp[-c_2\epsilon^2 n]$ for all n and ϵ.

The propositions above show that both the family $[Q_n]$ of cubes and the family $[S_n]$ of permutation graphs form normal Lévy families with parameter c_2 arbitrarily close to 2. In fact for the n–cube Q_n we can do slightly better – see the next subsection.

The notion of partition sequences above is quite a natural framework within which to apply the martingale inequalities (6.7) or (6.11), but it is not obvious that we have hit the 'right' level of generality. Milman & Schechtman (1986) give an attractive intermediate level which neatly contains examples (7.1) to (7.3). We shall consider only the finite case here.

Let G be a (finite) group with a translation invariant metric d; that is, $d(g,h) = d(rg,rh) = d(gr,hr)$ for all $g,h,r \in G$. Given a subgroup H we have a natural metric $\bar{d}$ defined on the set G/H of left cosets rH by setting $\bar{d}(rH,sH) = d(r,sH)$ where of course $d(r,sH)$ is the minimum value of $d(r,sh)$ over $h \in H$.

(7.9) Theorem: Let G be a group with a translation invariant metric d. Let $G = G_0 \supseteq G_1 \supseteq \cdots \supseteq G_n = \{e\}$ be a decreasing sequence of subgroups, and let c_k be the diameter of the space G_{k-1}/G_k for each $k = 1,\ldots,n$.

(a) Let the real–valued function f on G satisfy $|f(x)-f(y)| \le d(x,y)$ for all $x,y \in G$. Let X be uniformly distributed over G. Then for any $t > 0$

$$P(|f(X) - Ef(X)| \ge t) \le 2\exp\left[-2t^2/\Sigma c_k^2\right].$$

(b) Let A be a subset of G with $|A|/|G| = \alpha$, $0<\alpha<1$. Then for any $t \ge t_0 = \left[\frac{1}{2}\left[\log\frac{1}{\alpha}\right]\Sigma c_k^2\right]^{1/2}$,

$$|A_t|/|G| \ge 1 - \exp\left[-2\left[t-t_0\right]^2/\Sigma c_k^2\right];$$

and so for any $t > 0$ and $\gamma > 0$

$$|A_t|/|G| \ge 1 - \left[\frac{1}{\alpha}\right]^{\gamma^2} \exp\left[-2\left[\frac{\gamma}{1+\gamma}\right]^2 t^2/\Sigma c_k^2\right].$$

(b) Exact isoperimetric inequalities

A Hamming ball centred at a vertex v of the cube Q_n consists of all vertices at distance less than d from v for some d, together with some vertices at distance d.

Let A and C be two sets of vertices of Q_n with distance ρ between them. Then a seminal theorem of Harper (1966) (see also Bollobás (1986) page 129) asserts that there exist Hamming balls B_0 centred at the all–zero vector and B_1 centred at the all–one vector such that $|B_0| = |A|$, $|B_1| = |C|$ and the distance between B_0 and B_1 is at least ρ. This gives the exact solution to the isoperimetric inequality for the n–cube. From this result and a one–sided version of the Chernoff bounds (1.1) we may obtain the following result (see Amir & Milman (1980), Alon & Milman (1985)).

(7.10) Proposition: Let A be a set of 2^{n-1} vertices in the cube Q_n. Then for any $t \geq 0$,

$$|A_t|/2^n \geq 1 - \exp\left[-2t^2/n\right].$$

This result is 'cleaner' than the corresponding case of proposition (7.7). We now see that the graphs Q_n form a normal Lévy family with parameters $c_1 = 1$, $c_2 = 2$.

In the last subsection we obtained our isoperimetric inequality (7.7) for the cube from a result on concentration of measure. We may also reverse this process.

(7.11) Proposition: Let f be a function defined on the vertex set V of the n–cube Q_n such that if x and y are adjacent then $|f(x)-f(y)| \leq 1$. Let the random variable X be uniformly distributed over V. Let L be a median of f; that is, $P(f(X) \leq L) \geq \frac{1}{2}$ and $P(f(X) \geq L) \geq \frac{1}{2}$. Then for any $t > 0$

$$P(|f(X) - L| \geq t) \leq 2 \exp\left[-2t^2/n\right].$$

Proof: Let $A = \{x \epsilon V: f(x) \leq L\}$ and $B = \{x \epsilon V: f(x) \geq L\}$. Then

$$P(|f(X) - L| \leq t) \geq P\left[X \epsilon A_t \cap B_t\right]$$
$$= -1 + P\left[X \epsilon A_t\right] + P\left[X \epsilon B_t\right]$$
$$\geq 1 - 2\exp\left[-2t^2/n\right]$$

by proposition (7.10). □

(We already knew this inequality with L replaced by Ef(X), for example by lemma (1.2) with each $c_k = 1$. See also Milman & Schechtman (1986), page 142, for the relationship between deviations from the mean and from the median.)

Recently several further exact discrete isoperimetric inequalities have been obtained – see Bollobás & Leader (1988a), (1988b), (1988c) and subsection (d) below.

(c) Two results of Alon and Milman

Let H_k be a graph with vertex set V_k, for k=1,...,n. The cartesian product $G = \Pi H_k$ has vertex set ΠV_k, and two vertices $x = \left(x_1,...,x_n\right)$ and $y = \left(y_1,...,y_n\right)$ are adjacent if and only if they differ in exactly one coordinate and if this is the kth co–ordinate then x_k and y_k are adjacent in H_k. Note that G has diameter the sum of the diameters of the graphs H_k. From corollary (7.5) we deduce the following extension of proposition (7.7) about the n–cube.

(7.12) Proposition: For k = 1,...,n let H_k be a graph with diameter D_k, and let the graph G_n with vertex set V be the cartesian product ΠH_k. Let $A \subseteq V$ with $|A|/|V| = \alpha$, $0<\alpha<1$. Then for any $t \geq t_0 = \left[\frac{1}{2}\log\left[\frac{1}{\alpha}\right]\Sigma D_k^2\right]^{1/2}$,

$$|A_t|/|V| \geq 1 - \exp\left[-2\left(t-t_0\right)^2/\Sigma D_k^2\right];$$

and for any $t > 0$ and any $\gamma > 0$

$$|A_t|/|V| \geq 1 - \left[\frac{1}{\alpha}\right]^{\gamma^2}\exp\left[-2\left[\frac{\gamma}{1+\gamma}\right]^2 t^2/\Sigma D_k^2\right].$$

This result shows that if G_n is the Cartesian product of n copies of a fixed graph H then the G_n's form a normal Lévy family with parameter c_2 arbitrarily close to 2. In a remarkable paper, Alon & Milman (1985) used quite different methods to show that the family (G_n) is concentrated, and commented that the methods of Maurey and Schechtman which we are using here show that (G_n) is a normal Lévy family (though with $c_2 = 1/64$). Bollobás and Leader (1988c) show that we may also take $c_2 = 6D^2/(k^2-1)$ where k is the order of H.

Let us consider a second interesting result from Alon & Milman (1985). For $k \geq 2$, the odd graph 0_k has vertices corresponding to the (k–1)–subsets of a (2k–1)–set, and two vertices are adjacent if and only if the corresponding sets are disjoint. Thus 0_2 is the triangle K_3 and 0_3 is the Petersen graph; and the graph 0_k has diameter k–1 (see Biggs (1974)). It is shown in Alon & Milman (1985) that the family (0_k) is concentrated.

Now for $1 \leq k \leq n$ let $\hat{S}_{n,k}$ be the graph with vertices the k–subsets of an n–set, and with two vertices adjacent if and only if the corresponding sets A, B satisfy $|A \setminus B| = |B \setminus A| = 1$. Thus the graph $\hat{S}_{2k-1,k-1}$ has the same vertices as 0_k, and if two vertices are adjacent in this graph then they are at distance 2 in 0_k. Observe that $\hat{S}_{n,k}$ has diameter k if $k \leq n/2$.

(7.13) Proposition: Consider the graph $\hat{G} = \hat{S}_{n,k}$, with vertex $\hat{V}$ say, $|\hat{V}| = \binom{n}{k}$. Let $\hat{A} \subseteq \hat{V}$ with $|\hat{A}|/|\hat{V}| = \alpha$, $0<\alpha<1$. Then for any $t \geq t_0 = \left(\frac{1}{2}\left(\log\frac{1}{\alpha}\right)k\right)^{1/2}$,

$$|\hat{A}_t|/|\hat{V}| \geq 1 - \exp\left(-2\left(t-t_0\right)^2/k\right).$$

We may of course think of the vertices of the graph $\hat{S}_{n,k}$ as members of the vertex set $\{0,1\}^n$ of the n–cube Q_n. Then two vertices of $\hat{S}_{n,k}$ are adjacent in $\hat{S}_{n,k}$ if and only if they are at distance 2 in Q_n. Thus we see that the concentration phenomenon observed for the n–cube Q_n in proposition (7.7) holds also for the slice $\Sigma x_i = k$. From the above we find that the graphs 0_k form a normal Lévy family with parameter c_2 about 1/2.

Proof: Consider the graph $G = S_{n,k}$ in example (7.3), with vertex set V. For each vertex g in V let $\hat{g}$ be the k–set of elements listed in g. Let $A = \{g \epsilon V: \hat{g} \epsilon \hat{A}\}$. Then $|A|/|V| = |\hat{A}|/|\hat{V}|$. Also, for each vertex g in V

$$d_G(g,A) = d_{\hat{G}}(\hat{g},\hat{A}).$$

Hence, for any $t > 0$, $|A_t|/|V| = |\hat{A}_t|/|\hat{V}|$. Now we may complete the proof by applying corollary (7.5) to example (7.3). □

(d) Monotonic functions

We considered in proposition (7.7) above the uniform distribution over the vertex set $V = \{0,1\}^n$ of the n–cube Q_n. Now consider a more general distribution, of particular interest in the theory of random graphs (with n as the number of edges in a complete graph). Let $0<p<1$ and let the random variable X take values in V, with $P(X=x) = p^s(1-p)^{n-s}$ where $s = \Sigma x_k$.

Bollobás and Leader (1986b) give a beautiful exact isoperimetric inequality for down–sets (hereditary properties) in Q_n. From this they deduce various estimates which they note are much better than can be obtained from Azuma's inequality lemma (4.1), for the case when p (or 1–p) is very small. The martingale inequalities (6.5), (6.6) give an alternative approach.

(7.14) Lemma: Let f be a non–decreasing function defined on the vertex set V of the n–cube Q_n such that if x and y are adjacent then $|f(x) - f(y)| \leq 1$. Let the random variable X be distributed over V as above. Then, for any $0 \leq \epsilon \leq 1$,

$$P(f(X) - Ef(X) \geq \epsilon np) \leq \exp\left[-\tfrac{1}{3}\epsilon^2 np\right],$$
$$P(f(X) - Ef(X) \leq -\epsilon np) \leq \exp\left[-\tfrac{1}{2}\epsilon^2 np\right].$$

Proof: Let $\mathcal{F}_k$ be the σ–field generated by the natural partition $\mathcal{P}_k$ given in example (7.1), and let Y_k be $E[f(X)|\mathcal{F}_k]$. Then

$$-p \leq Y_k - Y_{k-1} \leq 1 - p,$$

and so we may use (6.5) and (6.6). □

(7.15) Proposition: Let the random variable X be distributed over the vertex set V of the n–cube Q_n as above; that is, $P(X=x) = p^s(1-p)^{n-s}$ where $s = \Sigma x_k$. Let $A \subseteq V$ be decreasing (that is, $x \in A$ and $y \leq x$ implies $y \in A$) with $P(X \in A) = \alpha$, $0<\alpha<1$. Let $t_0 = \left[2\log\left[\frac{1}{\alpha}\right]np\right]^{1/2}$. Then for $t_0 \leq t \leq t_0 + np$,

$$P\left[X \in A_t\right] \geq 1 - \exp\left[-\tfrac{1}{3}\left[t-t_0\right]^2/np\right].$$

Proof: Take $f(x) = d(x,A)$ in the lemma above. Note that $d(x,A) \leq \Sigma x_i$ since the all–zero vector is in A. Thus $t_1 = Ef(X)$ satisfies $t_1 \leq np$. By lemma (7.14)

$$\alpha = P(f(X) = 0) = P(f(X) \leq t_1 - t_1) \leq \exp\left[-\tfrac{1}{2}t_1^2/np\right],$$

and it follows that $t_1 \leq t_0$. But now by lemma (7.14) again, for $t_0 \leq t \leq t_0 + np$

$$\begin{aligned} 1 - P\left[X \in A_t\right] &= P(f(X) > t) \\ &\leq P\left[f(X) > t_1 + \left[t-t_0\right]\right] \\ &\leq \exp\left[-\tfrac{1}{3}\left[t-t_0\right]^2/np\right]. \end{aligned}$$

□

§8 Applications in operational research and computer science

(a) Bin packing

Given an n–vector $\underline{x} = \left[x_1,\ldots,x_n\right]$ where each $x_i \in [0,1]$, let $B(\underline{x})$ be the least number of unit size bins needed to store n items with these sizes. Let $X_1,\ldots,X_n$ be independent random variables each taking values in $[0,1]$. The bounded differences method lets us much strengthen previous work on the behaviour of the random variable $B = B\left[X_1,\ldots,X_n\right]$ (Grimmett (1985), Rhee (1985)).

(8.1) Theorem: For any $t > 0$

$$P(|B - E(B)| \geq t) \leq 2\exp\left[-2t^2/n\right].$$

This is the first application in Rhee & Talagrand (1987a) (with an improved bound). To prove it from lemma (1.2) we need only note that for $\underline{x}$, $\underline{x}' \in [0,1]^n$ we have $|B(\underline{x}) - B(\underline{x}')| \leq 1$ whenever $\underline{x}$ and $\underline{x}'$ differ in only one coordinate. Similar results hold for the number of bins used by certain heuristics – see Rhee & Talagrand (1987a).

Now let us use B_n in place of B above. From the subadditive inequality

$$E\left[B_{m+n}\right] \leq E\left[B_m\right] + E\left[B_n\right]$$

it follows that $\frac{1}{n}E\left[B_n\right] \to \beta$ as $n \to \infty$, where $\beta = \inf \frac{1}{n}E\left[B_n\right]$ $(0 \leq \beta \leq 1)$.

(8.2) Corollary: Let $\epsilon > 0$. Then

$$P\left[\left|\tfrac{1}{n}B_n - \beta\right| > \epsilon\right] = 0\left[\exp\left\{-(2-o(1))\epsilon^2 n\right\}\right].$$

(b) Knapsack problems

Let $\underline{b}$ be a fixed non–negative m–vector. Consider a list $\underline{x}_1 = \left(c_1, \underline{a}_1\right)$, ..., $\underline{x}_n = \left(c_n, \underline{a}_n\right)$ of (1+m)–vectors, where each $c_k \epsilon$ [0,1] and each m–vector $\underline{a}_k \geq \underline{0}$. Denote the list $\left(\underline{x}_1, \ldots, \underline{x}_n\right)$ by $\underline{x}$, and let $K(\underline{x})$ be the value of the corresponding 'multi–knapsack' problem

$$\max \Sigma c_k z_k$$

subject to

$$\Sigma \underline{a}_k z_k \leq \underline{b}$$
$$z_k = 0 \text{ or } 1 \qquad (k = 1, \ldots, n).$$

Now let $\underline{X}_1, \ldots, \underline{X}_n$ be independent random variables, where each $\underline{X}_k$ is a (1+m)–vector $\left(C_k, \underline{A}_k\right)$, with $C_k \epsilon$ [0,1] and the m–vector $\underline{A}_k \geq \underline{0}$. The behaviour of the corresponding random knapsack value $K = K\left(\underline{X}_1, \ldots, \underline{X}_n\right)$ has been investigated in Frieze & Clarke (1984), Meante et al (1984), Mamer & Schilling (1988), Schilling (1988).

As with bin packing we obtain a strong concentration result immediately from lemma (1.2). We need only note that if the lists $\underline{x}$, $\underline{x}'$ differ in only one coordinate vector $\left(c_k, \underline{a}_k\right)$ then $|K(\underline{x}) - K(\underline{x}')| \leq 1$, to obtain

(8.3) Theorem: For any $t > 0$,

$$P(|K - E(K)| \geq t) \leq 2\exp\left[-2t^2/n\right].$$

(c) Travelling salesman problem

Given n points $\underline{x}_1,\dots,\underline{x}_n$ in the unit square $[0,1]^2$ let $T\left[\underline{x}_1,\dots,\underline{x}_n\right]$ be the shortest length of a tour through them. Now let $\underline{X}_1,\dots,\underline{X}_n$ be independent random variables, each uniformly distributed on the unit square. Marvellous results are known related to the asymptotic behaviour of the random variable $T = T\left[\underline{X}_1,\dots,\underline{X}_n\right]$, starting with the seminal paper of Beardwood, Halton and Hammersley (1959) – see the survey article by Karp & Steele (1985). We are interested here in the concentration of T.

Given a fixed point $\underline{y}$ in the unit square let Y be the random shortest distance from $\underline{y}$ to one of the points $\underline{X}_{k+1},\dots,\underline{X}_n$. Then, as observed in Steele (1981), for some constant $c_1>0$

$$P(Y > t) \leq \left[1 - c_1t^2\right]^{n-k-1} \qquad (\text{for } t > 0).$$

Hence $E(Y) \leq c_2(n-k)^{-1/2}$ for some constant $c_2 > 0$. It follows by considering first $\underline{y} = \underline{x}_k$ then $\underline{y} = \underline{x}_k'$ that

$$|E\left[T|\left[\underline{X}_1,\dots,\underline{X}_k\right]=\left[\underline{x}_1,\dots,\underline{x}_k\right]\right] - E\left[T|\left[\underline{X}_1,\dots,\underline{X}_{k-1}\right]=\left[\underline{x}_1,\dots,\underline{x}_{k-1}\right], \underline{X}_k=\underline{x}_k'\right]|$$
$$\leq 2c_2(n-k)^{-1/2}.$$

So by Azuma's inequality lemma (4.1) we find that for $t > 0$

(8.4) $$P(|T - E(T)| \geq t) \leq \exp\left[-t^2/c\log n\right]$$

for a suitable constant $c > 0$. This result of Rhee & Talagrand (1987a) improves on work of Steele (1981) for large deviations t.

The inequality (8.4) is of no use if we are interested in small t, say $t = o\left[(\log n)^{1/2}\right]$. For such small deviations direct application of the bounded difference method fails – see Rhee & Talagrand (1987a), (1987b), (1989), Rhee (1988), Steele (1989).

(d) Minimum spanning trees

In the complete graph K_n with independent random edge lengths each uniformly distributed on [0,1], the expected length of a minimum spanning tree tends to $\zeta(3) \simeq 1.2$ as $n \to \infty$. This remarkable result is due to Frieze (1985). In Frieze & McDiarmid (1989) the result is extended and strengthened. This was possible partly because we could replace a messy second moment calculation by a simple use of the bounded differences method, and obtain a far stronger result.

Let us see how this goes. Let G be a fixed graph, with say n edges. Given a list $\underline{x} = \left[x_1,\ldots,x_n\right]$ of these edges, let $G_j(\underline{x})$ be the subgraph of G with edges $x_1,\ldots,x_j$, and let $\kappa_j(\underline{x})$ be the number of components of $G_j(\underline{x})$. Now suppose that the edges of G have independent random edge lengths each uniformly distributed on [0,1]. Let $\underline{X} = \left[X_1,\ldots,X_n\right]$ be the list of edges rearranged in increasing order. Thus all n! possible orders are equally likely. Knowing how fast $\kappa_j(\underline{X})$ usually decreases with j tells us how much of the minimum spanning tree we can expect to build with short edges.

It turns out then that we are interested in the random variable $Z = Z(\underline{X})$, where $Z(\underline{x}) = \Sigma\left\{\kappa_j(\underline{x}):\ j=1,\ldots,m\right\}$ and $m \leq n$. Here we focus on the concentration of Z. But as in example (7.3) we see that

$$\begin{aligned}&|E\left[\kappa_j(\underline{X})\,|\,\left[X_1,\ldots,X_k\right]=\left[e_1,\ldots,e_k\right]\right]\\ &\quad - E\left[\kappa_j(\underline{X})\,|\,\left[X_1,\ldots,X_{k-1}\right]=\left[e_1,\ldots,e_{k-1}\right],\ X_k=e_k'\right]| \leq 1.\end{aligned}$$

It now follows that

$$\begin{aligned}&|E(Z\,|\,\left[X_1,\ldots,X_{k-1}\right]=\left[e_1,\ldots,e_k\right])\\ &\quad -E(Z\,|\,\left[X_1,\ldots,X_{k-1}\right]=\left[e_1,\ldots,e_{k-1}\right],\ X_k=e_k'\right]| \leq m-k+1.\end{aligned}$$

Hence by corollary (6.10), for any $t > 0$

(8.5) $$P(|Z - E(Z)| \geq t) \leq 2\exp\left[-12t^2/m(m+1)(2m+1)\right].$$

(e) Second eigenvalue of random regular graphs

Let G be an r–regular graph, possibly with loops or multiple edges. The adjacency matrix (with loops counted twice) has r as the eigenvalue with largest absolute value. Let $\lambda_2(G)$ be the next largest eigenvalue in absolute value. If $|\lambda_2(G)|$ is much less than r then G has useful 'expansion' properties and the natural random walk on the vertices is 'rapidly mixing' – see for example Broder & Shamir (1987) and the references there. Both these properties are much sought in computer science.

Consider random 2d–regular graphs constructed as follows on the vertex set $V = \{1,...,n\}$. Let A be the set of all permutations σ on V. Pick $\sigma_1,...,\sigma_d$ independently at random from A, and let the graph G have the dn edges $\{v, \sigma_k v\}$ for $k = 1,...,d$ and $v \in V$. Let Y be the corresponding random variable $|\lambda_2(G)|$. We are interested in large n and moderate d. Broder & Shamir (1987) show essentially that $E[Y] \le (c + o(1))d^{3/4}$ as $n \to \infty$; and for any $t > 0$

$$P(|Y - E(Y)| \ge t) \le 2\exp\left[-t^2/8d\right]. \tag{8.6}$$

The concentration result (8.6) may be proved as follows. Standard manipulations show that

$$2d - \lambda_2 = \inf \Sigma_k \Sigma_v \left[f\left[\sigma_k v\right] - f(v)\right]^2,$$

where the infimum is over all real–valued functions f on V satisfying $\Sigma_v f(v) = 0$ and $\Sigma_v f(v)^2 = 1$. But if $\Sigma_v f(v)^2 = 1$ the triangle inequality gives

$$\Sigma_v \left[f\left[\sigma_k v\right] - f(v)\right]^2 \le 4.$$

Hence we may apply lemma (1.2) with each $c_k = 4$.

This neat application of the bounded differences method has been rather eclipsed by recent work of Friedman (1988) and Kahn & Szemerédi (1988), who show that Y concentrates near the lower bound of about $2(2d-1)^{1/2}$.

(f) Heap building

Suppose that we wish to build a heap on n elements, when all n! initial orders are equally likely (see for example Knuth (1973)). In McDiarmid & Reed (1989) a new algorithm is introduced, and it is shown that the random number B_n of comparisons required satisfies $\frac{1}{n}E\left[B_n\right] \to \alpha$ as $n \to \infty$, where $\alpha \simeq 1.52$. (This is the best known average case behaviour.) The result is given extra weight by the fact that B_n concentrates strongly around the mean. This concentration is our interest here.

The algorithm is a variant of Floyd's method (see Knuth (1973)) and consists of a sequence of 'merge' operations moving up a binary tree. We can use corollary (6.10), since the average effect on B_n of learning the history of one more merge is small. The reason for this is that as we move up the binary tree the expected number of times we revisit this old working is at most 1. We find the following result, and similar results for other variants of Floyd's method.

(8.7) Theorem For any $\epsilon > 0$, if n is sufficiently large

$$P\left[\left|\tfrac{1}{n}B_n - \alpha\right| \geq \epsilon\right] < \exp\left[-\left[\epsilon^2/9\right]n\right].$$

§9 Concluding Remarks

The bounded differences method is just the application of certain inequalities related to bounded martingale difference sequences, but we have seen that it is wonderfully useful in combinatorics and the mathematics of Operational Research. As long as there are small bounds and large deviations the inequalities really bite, and allow us to handle problems that seemed intractable only a short time ago.

How many more interesting applications will have appeared by the time of the conference in July?

Acknowledgement I would like to thank Martin Dyer for helpful comments.

References

Alon, N. & Milman, V.D. (1985). λ_1, isoperimetric inequalities for graphs, and superconcentrators. J. Comb. Th. B 38, 73–88.

Amir, D. & Milman, V.D. (1980). Unconditional and symmetric sets in n–dimensional normed spaces. Israel J. Math. 37, 3–20.

Angluin, D. & Valiant, L.G. (1979). Fast probabilistic algorithms for Hamiltonian circuits and matchings. J. Computer and System Sciences 18, 155–193.

Azuma, K. (1967). Weighted sums of certain dependent random variables. Tôkuku Math. J. 19, 357–367.

Bahadur, R.R. (1971). Some limit theorems in Statistics. Conference Board of the Mathematical Sciences Regional Conference Series in Applied Mathematics 4, SIAM, Philadelphia, USA.

Bahadur, R.R. & Ranga Rao, R. (1960). On deviations of the sample mean. Ann. Math. Statist. 31, 1015–1027.

Beardwood, J., Halton, J.H. and Hammersley, J. (1959). The shortest path through many points. Proc. Camb. Phil. Soc. 55, 299–327.

Biggs, N. (1974). Algebraic Graph Theory. Cambridge: Cambridge University Press.

Bollobás, B. (1985). Random Graphs. London: Academic Press.

Bollobás, B. (1986). Combinatorics. Cambridge: Cambridge University Press.

Bollobás, B. (1987). Martingales, isoperimetric inequalities and random graphs. Colloq. Math. Soc. János Bolyai 52, 113–139.

Bollobás, B. (1988a). The chromatic number of random graphs. Combinatorica 8, 49–55.

Bollobás, B. (1988b). Sharp concentration of measure phenomena in the theory of random graphs. manuscript.

Bollobás, B. & Erdős, P. (1976). Cliques in random graphs. Mathematical Proceedings of the Cambridge Philosophical Society 80, 419–427.

Bollobás, B. & Leader, I. (1988a). An isoperimetric inequality on the discrete torus. manuscript.

Bollobás, B. & Leader, I. (1988b). Isoperimetric inequalities and fractional set systems. manuscript.

Bollobás, B. & Leader, I. (1988c). Compressions and isoperimetric inequalities. manuscript.

Broder, A. & Shamir, E. (1987). On the second eigenvalue of random regular graphs. In 28th Annual Symposium on Foundations of Computer Science, 286–294.

Burkholder, D.L. (1973). Distribution function inequalities. Ann. Probab. 1, 19–42.

Chernoff, H. (1952). A measure of asymptotic efficiency for tests of a hypotheses based on the sum of observations. Ann. Math. Statist. 23, 493–507.

Chung, K.L. (1974). A course in probability theory, Second edition. New York and London: Academic Press.

Chvátal, V. (1979). The tail of the hypergeometric distribution. Discrete Math. 25, 285–287.

Doob, J.L. (1953). Stochastic Processes. New York: John Wiley and Sons.

Erdős, P. & Spencer, J. (1974). Probabilistic Methods in Combinatorics. New York: Academic Press.

Feller, W. (1968). An Introduction to Probability Theory and its Applications, Volume 1, Third Edition. New York: John Wiley and Sons.

Freedman, D.A. (1975). On tail probabilities for martingales. Ann. Probab. 3, 100–118.

Friedman, J. (1988). On the second eigenvalue and random walks in random d–regular graphs. manuscript.

Frieze, A.M. (1985). On the value of a random minimum spanning tree problem. Discrete Applied Math. 10, 47–56.

Frieze, A.M. (1989). On the independence number of random graphs. Discrete Math., to appear.

Frieze, A.M. & Clarke, M.R.B. (1984). Approximation algorithms for the m–dimensional 0–1 knapsack problem: worst case and probabilistic analysis. Europ. J. O. R. 15, 100–109.

Frieze, A.M. & Luczak, T. (1988). On the independence number of random regular graphs. manuscript.

Frieze, A.M. & McDiarmid, C.J.H. (1989). On random minimum length spanning trees. Combinatorica, to appear.

Garsia, A.M. (1973). Martingale inequalities, Seminar Notes on Recent Progresses. New York: W.A. Benjamin.

Gleser, L.J. (1975). On the distribution of the number of successes in independent trials. Ann. Probab. 3, 182–188.

Grimmett, G.R. (1985). Large deviations in subadditive processes and first–passage percolation. Contemporary Mathematics 41, 175–194.

Grimmett, G.R. & McDiarmid, C.J.H. (1975). On colouring random graphs. Mathematical Proceedings of the Cambridge Philosophical Society 77, 313–324.

Harper, L.H. (1966). Optimal numberings and isoperimetric problems on graphs. J. Comb. Th. 1, 385–393.

Hoeffding, W. (1956). On the distribution of the number of successes in independent trials. Ann. Math. Statist. 27, 713–721.

Hoeffding, W. (1963). Probability inequalities for sums of bounded random variables. J. Amer. Statist. Assoc. 58, 13–30.

Johnson, W.B., Schechtman, G. & Zinn, J. (1985). Best constants in moment inequalities for linear combinations of independent and exchangeable random variables. Ann. Prob. 13, 234–253.

Kahn, J. & Szemerèdi, E. (1988). manuscript.

Karp, R.M. (1979). A patching algorithm for the nonsymmetric travelling–salesman problem. SIAM J. Comput. 8, 561–573.

Karp, R.M., Lenstra, J.K., McDiarmid, C.J.H. & Rinnooy Kan, A.H.G. (1985). Probabilistic analysis of combinatorial algorithms: an annotated bibliography. In Combinatorial Optimisation: Annotated Bibliographies, edited by M. O'hEigeartaigh, J.K. Lenstra and A.H.G. Rinnooy Kan. Chichester, England: John Wiley and Sons.

Karp, R.M. & Steele, J.M. (1985). Probabilistic analysis of heuristics. In The Travelling Salesman Problem: A Guided Tour of Combinatorial Optimization, eds. E.L. Lawler et al. New York: John Wiley and Sons.

Kingman, J.F.C. (1976). Subadditive processes. Lecture Notes in Math. 539, 168–222. Berlin and New York: Springer–Verlag.

Knuth, D.E. (1973). The Art of Computer Programming, Volume III: Sorting and Searching. Reading, Mass: Addison–Wesley.

Korsunov, A.D. (1980). The chromatic number of n–vertex graphs. Metody Diskret. Analiz. No.35, 14–44, 104 (in Russian).

Luczak, T. (1988a). Note on the sharp concentration of the chromatic number of random graphs. manuscript.

Luczak, T. (1988b). The chromatic number of random graphs. manuscript.

McDiarmid, C.J.H. (1982). Achromatic numbers of random graphs. Math. Proc. Camb. Phil. Soc. 92, 21–28.

McDiarmid, C.J.H. (1984). Colouring random graphs. Ann. Oper. Res. 1, 183–200.

McDiarmid, C.J.H. (1989). On the chromatic number of random graphs. manuscript.

McDiarmid, C.J.H. (1989). Probabilistic analysis of tree search. manuscript.

McDiarmid, C.J.H. & Reed, B.A. (1989). Building heaps fast. J. Algorithms, to appear.

Mamer, J.W. & Schilling, K.E. (1988). On the growth of random knapsacks. manuscript.

Matula, D.W. (1970). On the complete subgraphs of a random graph. Combinatorical Mathematics and its Applications, Chapel Hill, 356–369.

Matula, D.W. (1972). The employee party problem. Notices A.M.S. 19, A–382.

Matula, D.W. (1976). The largest clique size in a random graph. Technical Report, Department of Computer Science, Southern Methodist University, Dallas, Texas.

Matula, D.W. (1987). Expose and merge exploration and the chromatic number of a random graph. Combinatorica 7, 275–284.

Maurey, B. (1979). Construction de suites symétriques. Compt. Rend. Acad. Sci. Paris 288, 679–681.

Meante, M., Rinnooy Kan, A.H.G., Stougie, L. & Vercellis, C. (1984). A probabilistic analysis of the multiknapsack value function. manuscript.

Milman, V. & Schechtman, G. (1986). Asymptotic theory of finite dimensional normed spaces. Lecture Notes in Math. 1200. Berlin and New York: Springer–Verlag.

Rhee, W.T. (1985). Convergence of optimal stochastic bin packing. Operations Research Letters 4, 121–123.

Rhee, W.T. (1988). On the fluctuations of the stochastic travelling salesperson problem. manuscript, Ohio State University.

Rhee, W.T. & Talagrand, M. (1987a). Martingale inequalities and NP–complete problems. Math. of O. R. 12, 177–181.

Rhee, W.T. & Talagrand, M. (1987b). Martingale inequalities, interpolation and NP–complete problems. manuscript.

Rhee, W.T. & Talagrand, M. (1987c). Martingale inequalities and the jacknife estimate of variance. manuscript.

Rhee, W.T. & Talagrand, M. (1989). A sharp deviation inequality for the stochastic travelling salesman problem. Ann. Probab. 17, 1–8.

Schechtman, G. (1982). Lévy type inequality for a class of finite metric spaces. Lecture Notes in Math. 939, 211–215. Berlin and New York: Springer–Verlag.

Scheinerman, E.R. (1989). On the interval number of random graphs. Discrete. Math., to appear.

Schilling, K.E. (1988). On the growth of m–dimensional random knapsacks. manuscript.

Shamir, E. (1988a). A slightly random source confronts a random witness–set. manuscript

Shamir, E. (1988b). Chromatic numbers of random hypergraphs and associated graphs. In Randomness and Computation, ed. S. Micali. Greenwich, Connecticut: JAI Press.

Shamir, E. (1988c). Generalised stability and chromatic numbers of random graphs. manuscript.

Shamir, E. & Spencer, J. (1987). Sharp concentration of the chromatic number on random graphs $G_{n,p}$. Combinatorica 7, 121–129.

Steele, J.M. (1981). Complete convergence of short paths and Karp's algorithm for the TSP. Math. Oper. Res. 6, 374–378.

Steele, J.M. (1989). Seedlings in the theory of shortest paths. manuscript.

Steiger, W.L. (1967). Some Kolmogoroff–type inequalities for bounded random variables. Biometrica 54, 641–648.

Steiger, W.L. (1969). A best possible Kolmogoroff–type inequality for martingales and a characteristic property. Ann. Math. Statistics 40, 764–769.

Steiger, W.L. (1970). Bernstein's Inequality for Martingales. Z. Wahrscheinlichkeitstheorie verw. Geb. 16, 104–106.

Stout, W.F. (1974). Almost Sure Convergence. New York: Academic Press (lemma 4–2–3 and exercise 4–2–2).

Institute of Economics and Statistics,
Oxford University,
St. Cross Building,
Manor Road,
Oxford,
OX1 3UL.

ON THE USE OF REGULAR ARRAYS IN THE CONSTRUCTION OF t-DESIGNS

L. Teirlinck

1. INTRODUCTION

A t-design $S(\lambda; t, k, v)$ is a collection of k-subsets, called blocks, of a v-set S such that any t-subset of S is contained in exactly λ blocks. An $S(\lambda; 2, k, v)$ is often called a (v, k, λ)-design and an $S(\lambda; t, k, v)$ is often called a t-(v, k, λ)-design. An $S(\lambda; t, k, v)$ is called simple if it contains no repeated blocks. It has been known for a long time that there are a lot of $S(\lambda; t, k, v)$ for all t, see [6, 23]. However, until relatively recently, the only known examples of simple t-designs with $t \geq 6$ were the trivial t-designs consisting of all k-subsets of a v-set. The first examples of non-trivial simple 6-designs were found by Magliveras and Leavitt [9]. In [19], we constructed nontrivial simple t-designs for all t. It is not the purpose of this paper to give another proof of the main result of [19], as a simplified proof is already given in [21]. Rather, we will survey construction techniques for t-designs using totally symmetric regular arrays, or, equivalently, regular extended designs. These techniques played a major role in the construction of non-trivial simple t-designs for arbitrary t. We will point out the relationship between the techniques of [19, 21] and results of Wilson, Schreiber, Beth and Lu, as well as other results of the author and folklore direct product constructions. Another major element of the constructions of [19, 21], namely what are called $LS(\lambda; t, k, \delta)$ in [21], will be generalized in a subsequent publication [22], in the context of t-designs on posets.

2. PRELIMINARY DEFINITIONS AND RESULTS.

Throughout this paper, we assume that sets are finite unless they are infinite for obvious reasons. This is mainly a matter of convenience. Most constructions do not require finiteness. However, nearly all of the structures we will consider are very easy to construct in the infinite case, using straightforward inductive limit techniques. (See for instance [22].) This makes more sophisticated constructions less relevant for infinite structures.

An X-multiset, or simply multiset, will be a function $\mu: X \to \mathbb{N}$. We call $\mu(x)$ the multiplicity of x. We usually view a multiset as a set with repeated elements. We call x an element of μ if $\mu(x) \neq 0$ and a repeated element if $\mu(x) \geq 2$. We consider an X-multiset μ to be also a Y-multiset for all $Y \supset X$ by using the convention that $\mu(y) = 0$ for all $y \notin X$. We often describe a multiset by a collection of elements between square brackets. For instance, [a, a, b, c, c, c] denotes the {a, b, c}-multiset defined by $\mu(a) = 2$, $\mu(b) = 1$ and $\mu(c) = 3$. The square brackets are used to avoid confusion with ordered or unordered sets. A multiset is called simple if it has no repeated elements. We do not distinguish between sets and simple multisets. For instance, if A and B are sets and if $\lambda_1, \lambda_2 \in \mathbb{N}$, then $\lambda_1 A + \lambda_2 B$ denotes the $(A \cup B)$-multiset containing λ_1 copies of each element of $A - B$, λ_2 copies of each element of $B - A$ and $\lambda_1 + \lambda_2$ copies of each element of $A \cap B$. A t-X-multiset, or simply t-multiset, will be an X-multiset μ such that $|\mu| = \sum_{x \in X} \mu(x) = t$. If μ is a multiset, we denote by $s(\mu)$ the underlying set of μ, i.e. $s(\mu) = \{x;\ x \in \mu\}$. By the number of elements of μ having a given property, we always mean the sum of the multiplicities of the elements of $s(\mu)$ having that property. If X is a set, then P(X) is the set of all subsets of X, $P_k(X)$ the set of all k-subsets and $P_{k_1,k_2}(X)$ the set of all $B \in P(X)$ with $k_1 \leq |B| \leq k_2$. If μ is a P(X)-multiset, we often call the elements of X points and the elements of μ blocks. A t-design $S(\lambda;\ t, k, v)$, $\lambda, t, k, v \in \mathbb{N}$, $t \leq k$, can be viewed as a $P_k(S)$-multiset μ, $|S| = v$, such that every t-subset of S is contained in exactly λ elements of μ. For t-designs and related notions, we use the convention that if λ is not specified, we have $\lambda = 1$. Thus, we often write S(t, k, v) instead of S(1; t, k, v). We sometimes write the "1", for instance to emphasize that a particular remark only applies to the case $\lambda = 1$. A well known necessary condition for the existence of an $S(\lambda; t, k, v)$, $v \geq k$, $t > 0$, is that $\lambda \cdot \binom{v-i}{t-i}/\binom{k-i}{t-i}$ should be an integer for all $i = 0, 1, \ldots, t-1$. If $k = t+1$, this simplifies to the condition that λ should be divisible by $\lambda(t, t+1, v) = \gcd(v-t, \ell cm \{1, \ldots, t+1\})$.

A <u>large set of disjoint</u> $S(\lambda; t, k, v)$, briefly $LS(\lambda; t, k, v)$, is a collection $(\mu_r)_{r\varepsilon R}$ of $S(\lambda; t, k, v)$ on a v-set S, $v \geq k$, such that $\sum_{r\varepsilon R} \mu_r = P_k(S)$. In other words, an $LS(\lambda; t, k, v)$ is a partition of $P_k(S)$ into $S(\lambda; t, k, v)$. Obviously we must have $|R| = \binom{v-t}{k-t}/\lambda$. If $k = t+1$, we have $|R| = (v-t)/\lambda$.

If S is a set, then $P_k(S)$ is an $S(\binom{v-t}{k-t}; t, k, v)$. An $S(\lambda, t, k, v)$ design μ on a set S is called <u>trivial</u> if $\binom{v-t}{k-t}$ divides λ and $\mu = (\lambda/\binom{v-t}{k-t})\, P_k(S)$. An $LS(\lambda; t, k, v)(\mu_r)_{r\varepsilon R}$ is called <u>trivial</u> if $\lambda = \binom{v-t}{k-t}$. (This implies that R consists of a single element r and that $\mu_r = P_k(S)$.) Obviously, if $(\mu_r)_{r\varepsilon R}$ is a non-trivial $LS(\lambda; t, k, v)$, then, for every $r \varepsilon R$, μ_r is a non trivial simple $S(\lambda; t, k, v)$.

For a review of some known results on $S(\lambda; t, k, v)$ we refer, for instance, to [3]. For a review of known results on $LS(\lambda; t, k, v)$ we refer to [20] for $\lambda=1$ and to [21] for $k = t+1$ and arbitrary λ.

If S is a set and $k \varepsilon N$, then, as usual, S^k denotes the set of all k-tuples $(x_1, \ldots, x_k)$, where $\{x_1, \ldots, x_k\} \subset S$. A <u>k-S-array</u> will be an S^k-multiset. The elements of the array are called <u>rows</u>. If we do not want to specify k or S, we use terms such as <u>k-array</u>, <u>S-array</u> or <u>array</u>. To avoid multiple brackets, we often write $\mu(x_1, \ldots, x_k)$ instead of $\mu((x_1, \ldots, x_k))$. A k-S-array μ is called <u>totally symmetric</u> if, for any $\sigma \varepsilon \mathcal{S}_k$, we have $\mu(x_1, \ldots, x_k) = \mu(x_{\sigma(1)}, \ldots, x_{\sigma(k)})$. If S is a set and $k \varepsilon N$, then $M(S)$ will denote the set of all S-multisets and $M_k(S)$ will denote the set of all k-S-multisets. If μ is a totally symmetric k-S-array, then we can define an $M_k(S)$-multiset μ' by putting $\mu'([x_1, \ldots, x_k]) = \mu(x_1, \ldots, x_k)$. (Note that, as μ is totally symmetric, $\mu(x_1, \ldots, x_k)$ does not depend on the order in which we write $x_1, \ldots, x_k$.) Obviously, the mapping g defined by $g(\mu) = \mu'$ is a bijection between the set of all totally symmetric k-S-arrays and the set of all $M_k(S)$-multisets.

An <u>orthogonal array</u> $OA(\lambda; t, k, v)$ on a v-set S is a k-S-array μ such that for any $\{i_1, \ldots, i_t\} \in P_t(\{1, \ldots, k\})$ and for any t-tuple $x_{i_1}, \ldots, x_{i_t}$ of (not necessarily distinct) elements of S, there are exactly λ rows $(y_1, \ldots, y_k)$ of μ with $y_{i_1} = x_{i_1}, \ldots, y_{i_t} = x_{i_t}$. A <u>large set of disjoint</u> $OA(\lambda; t, k, v)$, briefly $LA(\lambda; t, k, v)$, on a v-set S, is a collection $(\mu_r)_{r\in R}$ of $OA(\lambda; t, k, v)$ on S such that $\sum_{r\in R} \mu_r = S^k$. Obviously, we must have $|R| = v^{k-t}/\lambda$. An $LA(\lambda; t, k, v)(\mu_r)_{r\in R}$ is called totally symmetric and denoted by $LSA(\lambda; t, k, v)$ if all μ_r are totally symmetric.

We say that a multiset μ <u>contains</u> a multiset ν and write $\nu \leq \mu$ if $s(\nu) \subset s(\mu)$ and $\nu(x) \leq \mu(x)$ for all $x \in s(\nu)$. An <u>extended design</u> $ES(\lambda; t, k, v)$ on a v-set S is an $M_k(S)$-multiset μ such that every element of $M_t(S)$ is contained in exactly λ elements of μ. (Note that we do not consider multiple containment. For instance, if μ is an $M_3(S)$-multiset and if $x \in S$, then in counting the number of blocks of μ containing $[x, x]$, we do <u>not</u> count blocks of the type $[x, x, x]$ three times each.) Extended designs were first studied, for $t = 2$ and $k = 3$, by E. H. Moore [10], but the name extended design was given by Johnson and Mendelsohn [7]. Extended designs are one of several possible generalizations of designs that replace sets by multisets. For a review of other notions, we refer to [4]. A <u>large set of disjoint</u> $ES(\lambda; t, k, v)$, briefly $LES(\lambda; t, k, v)$, is a collection $(\mu_r)_{r\in R}$ of $ES(\lambda; t, k, v)$ on a v-set S such that $\sum_{r\in R} \mu_r = M_k(S)$. Obviously, we must have $|R| = |M_{k-t}(S)|/\lambda = \binom{k-t+v-1}{k-t}/\lambda$. It is easy to see that a totally symmetric $(t+1)$-array μ is an $OA(\lambda; t, t+1, v)$ iff μ' is an $ES(\lambda; t, t+1, v)$. Moreover, we have $(S^k)' = M_k(S)$ and, for any collection $(\mu_r)_{r\in R}$ of totally symmetric k-S-arrays, we have $(\sum_{r\in R} \mu_r)' = \sum_{r\in R} \mu_r'$. Thus, there is a one-to-one correspondence between totally symmetric $OA(\lambda; t, t+1, v)$ ($LSA(\lambda; t, t+1, v)$, respectively) on a v-set S and $ES(\lambda; t, t+1, v)$ ($LES(\lambda; t, t+1, v)$, respectively) on S. On the other hand, it is easy to check that totally symmetric $OA(1; t, k, v)$ with $k > t+1$ and $t \geq 2$, can only exist for $v \in \{0, 1\}$. Thus $ES(1; t, k, v)$

(LES(1; t, k, v), respectively) seem to be the most natural generalization of totally symmetric OA(1; t, t+1, v)(LSA(1; t, t+1, v), respectively) to the case $k > t+1$.

A k-S-array μ is called <u>$(\lambda; n)$-regular</u>, $0 \leq n \leq k-1$, if for any $\{i_1, \ldots, i_n\} \in P_n(\{1, \ldots, k\})$ and for any n-tuple $x_{i_1}, \ldots, x_{i_n}$ of elements of S, there are exactly λ rows $(y_1, \ldots, y_k)$ of μ such that $x_{i_1} = y_{i_1}, \ldots, x_{i_n} = y_{i_n}$ and $y_i = y_j$ for all $i, j \in \{1, \ldots, k\} - \{i_1, \ldots, i_n\}$. An $M_k(S)$-multiset ν is called <u>$(\lambda;n)$-regular</u>, $0 \leq n \leq k-1$, if for any $C \in M_n(S)$, there are exactly λ elements B of ν of the form $C + (k-n)\{x\}$, $x \in S$. That is, there are exactly λ elements of ν that can be obtained by completing C into a k-S-multiset by adding $(k-n)$ copies of a single element of S. (When studying M(S)-multisets, we often denote elements of M(S) by capitals instead of greek letters. This is done to avoid confusion between S-multisets and M(S)-multisets. Also, we want to use the same type of notation, namely capitals, for blocks of P(S)-multisets and for blocks of M(S)-multisets.) Obviously, a totally symmetric k-S-array μ is $(\lambda; n)$-regular iff μ' is $(\lambda; n)$-regular. A k-S-array μ is $(\lambda; k-1)$-regular iff it is an OA(λ; k-1, k, $|S|$). An $M_k(S)$-multiset ν is $(\lambda; k-1)$-regular iff ν is an ES(λ; k-1, k, $|S|$). An OA(λ; t, t+1, v) or ES(λ; t, t+1, v) is called <u>regular</u> if it is $(\lambda; n)$-regular for all n with $0 \leq n \leq t$. A collection $(\mu_r)_{r \in R}$ of k-S-arrays ($M_k(S)$-multisets) is called $(\lambda; n)$-regular (regular) if all μ_r are $(\lambda; n)$-regular (regular).

If (G, +) is an abelian group and if $a \in G$, then $\gamma_{(t,(G,+),a)}$ will denote the set of all $(x_1, \ldots, x_{t+1}) \in G^{t+1}$ such that $x_1 + x_2 + \ldots + x_{t+1} = a$. If no confusion is possible, we write $\gamma_{(t, G, a)}$ instead of $\gamma_{(t,(G,+), a)}$. The following proposition is straightforward. (A proof can be found in [18].)

<u>Proposition 1</u>: (1) If (G, +) is an abelian group and if $t \in N$, then $\{\gamma_{(t,G,a)}; a \in G\}$ is an LSA(1; t, t+1, $|G|$) which is (1; n)-regular for every $0 \leq n \leq t$ such that $\gcd(t+1-n, |G|) = 1$.
(2) If $v \equiv 0 \pmod{\lambda}$, then $\{\gamma_{(t,(Z_v,+), \lambda a)} \cup \gamma_{(t,(Z_v,+), \lambda a+1)} \cup$

$\ldots \bigcup \gamma_{(t,(Z_v,+),\ \lambda a+\lambda-1)}$; $a = 0, 1, \ldots, (v/\lambda)-1\}$ is an LSA(λ; t, t+1, v) which is (λ; n)-regular for every $0 \leq n \leq t$ such that gcd(t+1-n, v) divides λ.

Note that Proposition 1 (1) shows that LSA(t, t+1, v) and thus LES(t, t+1, v) exist for all t and v. For a survey of some known results about OA(2, k, v) with k > 3, we refer to [3]. For a survey of known results about OA(t, k, v) with t > 2 and k > t+1, we refer to [11]. An LA(t, k, v) exists iff an OA(t, k, v) exists [20]. An ES(λ; 2, k, v) on a set S is called idempotent if $\mu(k\{x\}) = \lambda$ for all $x \in S$. An ES(λ; 2, k, v) on a set S is called totally unipotent with unit 1, $1 \in S$, if $\mu((k-1)\{x\} + \{1\}) = \lambda$ for all $x \in S$. If S is a v-set and $1 \notin S$, then there are obvious one-to-one correspondences between the set of all idempotent ES(λ; 2, k, v) on S, the set of all totally unipotent ES(λ; 2, k, v+1) with unit 1 on $S \cup \{1\}$ and the set of all S(λ; 2, k, v) on S. Some constructions for S(λ; t, k, v) have obvious analogues for ES(λ; t, k, v). However, to the author's best knowledge, the only pair (t, k) with k > t+1 and $t \geq 2$ for which the existence problem for ES(1; t, k, v) has been completely solved is (t, k) = (2, 4). It is proved in [1] that an ES(2, 4, v) exists iff $v \notin \{6, 8, 9\}$. We do not know any non-trivial results about LES(t, k, v) with k > t+1 and $t \geq 2$.

3. COMPLETING A REGULAR (L)ES(λ; t, t+1, v) to an (L)S(λ; t, t+1, v+t)

A regular ES(λ; t, t+1, v) looks a lot like an S(λ; t, t+1, v+t) from which t points have been deleted. (This statement can be made more precise, at the expense of getting rather technical. See [21, Proposition 3].) Also, it is well known that there is a bijection between the set of all (t+1)-subsets of a (v+t)-set and the set of all (t+1)-multisets of a v-set. Moreover, by Proposition 1 (2), there is a regular LSA(λ(t, t+1, v+t); t, t+1, v), and thus a regular LES(λ(t, t+1, v+t); t, t+1, v) for all t and v, where λ(t, t+1, v) is defined as in section 2. The thing that immediately comes to mind is to take a regular (L)ES(λ; t, t+1, v) on a v-set S, add a t-set $\{\infty_1, \ldots, \infty_t\}$ disjoint from S, and try to construct an (L)$S(\lambda; t, t+1, v+t)$ on $S \cup \{\infty_1, \ldots, \infty_t\}$. For $t=1$, this just leads to some trivial graph theoretic results.

Before looking at the case $t \geq 2$, let us formulate our problem more precisely. If μ is an $M(S)$-multiset, let μ^* be the $P(S)$-multiset obtained by replacing all multisets in μ by their underlying set. If μ is a $P(S)$-multiset and $Y \subset S$, we denote by $\mu \| Y$ the $P(Y)$-multiset obtained by replacing every B in μ by $B \cap Y$. Given a regular $ES(\lambda; t, t+1, v)\mu$ on a set S and a t-set $\{\infty_1, \ldots, \infty_t\}$ disjoint from S, we want to find an $S(\lambda; t, t+1, v+t)\nu$ on $S \cup \{\infty_1, \ldots, \infty_t\}$ such that $\nu \| S = \mu^*$.

We first look at the case $t = 2$ and $\lambda = 1$. If μ is a regular $ES(2, 3, v)$ on a v-set S, then μ^* contains exactly one singleton $\{x\}$ and the 2-sets in μ^* form a graph on $S-\{x\}$, consisting of a disjoint union of cycles of length ≥ 3. If $\{\infty_1, \infty_2\} \cap S = \emptyset$, $\infty_1 \neq \infty_2$, then there is an $S(2, 3, v+2)\nu$ on $S \cup \{\infty_1, \infty_2\}$ such that $\nu \| S = \mu^*$ iff all these cycles have even length. The corresponding problem for large sets is more complicated. If $(\mu_r)_{r\in R}$ is a regular $LES(2, 3, v)$ on S, then even if for every μ_r we can construct an $S(2, 3, v+2)\nu_r$ on $S \cup \{\infty_1, \infty_2\}$ such that $\nu_r \| S = \mu_r^*$, this does not guarantee that there is an $LS(2, 3, v+2)$ $(\varepsilon_r)_{r\in R}$ on $S \cup \{\infty_1, \infty_2\}$ such that $\varepsilon_r \| S = \mu_r^*$ for all $r\in R$.

The most obvious candidates for constructing $S(2, 3, v+2)$ or $LS(2, 3, v+2)$ in this way are the regular $LES(2, 3, v)$ $(\gamma'_{(2,G,a)})_{a\in G}$, where G is an abelian group of order v with $\gcd(v, 6) = 1$. This problem was studied independently by Schreiber [13] and Wilson [24]. Let G be an abelian group of order v, $\gcd(v, 6) = 1$, and let $\{\infty_1, \infty_2\} \cap G = \emptyset$, $\infty_1 \neq \infty_2$. Wilson and Schreiber showed that there is an $S(2, 3, v+2)\nu_a$ on $G \cup \{\infty_1, \infty_2\}$ with $\nu_a \| G = \gamma'^*_{(2,G,a)}$ iff the order of -2 modulo p is even for all prime divisors p of v. If β_0 is an $S(2, 3, v+2)$ on $G \cup \{\infty_1, \infty_2\}$ such that $\beta_0 \| G = \gamma'^*_{(2,G,0)}$, then we can put $\beta_a = \{\{x+a, y+a, z+a\};\ \{x,y,z\} \in \beta_0\}$, where we use the convention that $\infty_1 + a = \infty_1$ and $\infty_2 + a = \infty_2$. Clearly, $\beta_a \| G = \gamma'^*_{(2,G,3a)}$. Schreiber and Wilson proved that $(\beta_a)_{a\in G}$ is an $LS(2, 3, v+2)$ iff the order of -2 modulo p is congruent to $2 \pmod 4$ for all prime divisors p of v. (Note that if p is a prime, then a quadratic reciprocity argument shows that $p \equiv 7 \pmod 8$ is a sufficient condition for the order -2 modulo p to be congruent to $2 \pmod 4$.)

Let us next look at the case $t=3$. Inspired by the results of Schreiber and Wilson for $t=2$, the following approach seems natural. Let G be an abelian group of order v with $\gcd(v, 6) = 1$ and let $\{\infty_1, \infty_2, \infty_3\} \cap G = \emptyset$, $|\{\infty_1, \infty_2, \infty_3\}| = 3$. We could try to find an $S(3, 4, v+3)\beta_0$ on $G \cup \{\infty_1, \infty_2, \infty_3\}$ such that $\beta_0 \| G = \gamma'^{*}_{(3,G,0)}$, define $\beta_a = \{\{x+a, y+a, z+a, w+a\};\ \{x, y, z, w\} \in \beta_0\}$, where $\infty_i + a = \infty_i$, and hope to obtain an $LS(3, 4, v+3)$ by this procedure. However, it was proved by Beth [2], that although in some cases it is possible to construct an $S(3, 4, v+3)\beta_0$ on $G \cup \{\infty_1, \infty_2, \infty_3\}$ with $\beta_0 \| G = \gamma'^{*}_{(3,G,0)}$, the corresponding collection $(\beta_a)_{a\in G}$ is never an $LS(3, 4, v+3)$. In [2], some isolated examples of abelian groups G for which there is an $S(3, 4, v+3)\beta_0$ on $G \cup \{\infty_1, \infty_2, \infty_3\}$ with $\beta_0 \| G = \gamma'^{*}_{(3,G,0)}$ are given. However, the problem of deciding for which abelian groups G, $\gcd(|G|,6) = 1$, such a β_0 exists seems complicated. For $t > 3$, the problem becomes even more difficult.

Instead of using all the blocks of the $ES(t, t+1, v)\gamma'_{(t,G,a)}$, one could try to first throw away some blocks and then try to complete the remaining multiset to an $S(t, t+1, v+t)$. Wilson [24] used this approach to give a single direct construction for $S(2, 3, n)$ working for all $n \equiv 1$ or $3 \pmod 6$. Moreover, for $n \equiv 5 \pmod 6$, he obtained, in a similar way, an $S(2, \{3, 5\}, n)$ having exactly one block of size 5. (If $K \subset \mathbb{N}$, then an $S(\lambda; t, K, v)$ is a $P(S)$-multiset μ, S a v-set, such that any t-subset of S is contained in exactly λ blocks and such that $|B| \in K$ for all $B\in\mu$. In particular, an $S(\lambda; t, \{k\}, v)$ is the same thing as an $S(\lambda; t, k, v)$.) Whether a similar approach could yield useful results for $t \geq 3$ is an interesting, but difficult, question.

An <u>ordered design</u> $OD(\lambda; t, k, v)$ on a v-set S is a k-S-array μ such that any row of μ contains k distinct elements and such that, for any $\{i_1, \ldots, i_t\} \in P_t(\{1, \ldots, k\})$ and for any t-tuple $x_{i_1}, \ldots, x_{i_t}$ of distinct elements of S, there are exactly λ rows $(y_1, \ldots, y_k)$ of μ with $y_{i_1} = x_{i_1}, \ldots, y_{i_t} = x_{i_t}$. Put

$DS^k = \{(x_1, \ldots, x_k) \in S^k;\ |\{x_1, \ldots, x_k\}| = k\}$. A <u>large set of disjoint</u> $OD(\lambda; t, k, v)$, briefly $LD(\lambda; t, k, v)$, on a v-set S is a collection $(\mu_r)_{r\in R}$ of $OD(\lambda; t, k, v)$ on S such that $\sum_{r\in R} \mu_r = DS^k$. The type of problem studied in this section is a lot easier in the ordered case than in the unordered case. Indeed, it is proved in [20] that if μ is any regular $OA(\lambda; t, t+1, v)$ on a set S and if $\{\infty_1, \ldots \infty_t\}$ is a t-set disjoint from S, then one can always construct an $OD(\lambda; t, t+1, v+t)\nu$ on $S \cup \{\infty_1, \ldots, \infty_t\}$, which is related to μ in a way which is an ordered analogue of the condition $\nu \| S = \mu^*$. (We could make the last part of the preceding statement more precise, but this would get rather technical.) If $(\mu_r)_{r\in R}$ is a regular $LA(\lambda; t, t+1, v)$ on S, then the construction in [20] yields an $LD(\lambda; t, t+1, v+t)(\nu_r)_{r\in R}$ on $S \cup \{\infty_1, \ldots, \infty_t\}$.

4. DIRECT PRODUCTS

If μ_1 is a k-S_1-array and μ_2 is a k-S_2-array, then the <u>direct product</u> $\mu_1 \times \mu_2$ of μ_1 and μ_2 is the k-$(S_1 \times S_2)$-array defined by $(\mu_1\times\mu_2)((x_1,y_1), \ldots, (x_k,y_k)) = \mu_1(x_1, \ldots, x_k) \cdot \mu_2(y_1, \ldots, y_k)$. If μ_1 and μ_2 are totally symmetric, then so is $\mu_1\times\mu_2$. As $M_k(S)$-multisets are in one-to-one correspondence with totally symmetric k-S-arrays, this means that we have implicitly defined a direct product for $M_k(S)$-multisets. An explicit retranslation in terms of $M_k(S)$-multisets yields the following. If ν_1 is an $M_k(S_1)$-multiset and ν_2 is an $M_k(S_2)$-multiset, then $\nu_1\times\nu_2$ is the $M_k(S_1\times S_2)$-multiset defined by $(\nu_1\times\nu_2)([(x_1, y_1), \ldots, (x_k, y_k)]) = \nu_1([x_1, \ldots, x_k]) \cdot \nu_2([y_1, \ldots, y_k])$. If μ_1 and μ_2 are totally symmetric k-S_i-arrays, $i = 1, 2$, then $(\mu_1\times\mu_2)' = \mu_1'\times\mu_2'$. If μ_1 is an $OA(\lambda_1; t, k, v_1)$ and μ_2 is an $OA(\lambda_2; t, k, v_2)$, then $\mu_1\times\mu_2$ is an $OA(\lambda_1\lambda_2; t, k, v_1v_2)$. If $(\mu_a)_{a\in A}$ is an $LA(\lambda_1; t, k, v_1)$ and $(\mu_b)_{b\in B}$ is an $LA(\lambda_2; t, k, v_2)$, then $(\mu_a\times\mu_b)_{(a,b)\in A\times B}$ is an $LA(\lambda_1\lambda_2; t, k, v_1v_2)$. If μ_1 is an $ES(\lambda_1; t, t+1, v_1)$ and μ_2 is an $ES(\lambda_2; t, t+1, v_2)$, then $\mu_1\times\mu_2$ is an $ES(\lambda_1\lambda_2; t, t+1, v_1v_2)$. If $(\mu_a)_{a\in A}$ is an $LES(\lambda_1; t, t+1, v_1)$ and $(\mu_b)_{b\in B}$ is an $LES(\lambda_2; t, t+1, v_2)$, then $(\mu_a\times\mu_b)_{(a,b)\in A\times B}$ is an $LES(\lambda_1\lambda_2; t, t+1, v_1v_2)$. However, if μ_1 is an $ES(\lambda_1; t, k, v_1)$ and μ_2 is an $ES(\lambda_2; t, k, v_2)$, $k > t+1$, then nothing guarantees that

$\mu_1 \times \mu_2$ will be an $ES(\lambda;\ t,\ k,\ v_1v_2)$ for any value of λ.

We have seen, in Section 2, that $S(\lambda;\ 2,\ 3,\ v)$ on a set S are in one-to-one correspondence with idempotent $ES(\lambda;\ 2,\ 3,\ v)$ on S and totally unipotent $ES(\lambda;\ 2,\ 3,\ v+1)$ with unit 1 on $S \cup \{1\}$, $1 \notin S$. The direct product of two idempotent (totally unipotent) $ES(\lambda_i;\ 2,\ 3,\ v_i)$, $i=1,\ 2$, is an idempotent (totally unipotent) $ES(\lambda_1\lambda_2;\ 2,\ 3,\ v_1v_2)$. Thus, the direct product described above, implicitly defines two direct product constructions for $S(\lambda;\ 2,\ 3,\ v)$, called the squag product and the sloop product. Starting with an $S(\lambda_1;\ 2,\ 3,\ v_1)$ and an $S(\lambda_2;\ 2,\ 3,\ v_2)$, the squag product yields an $S(\lambda_1\lambda_2;\ 2,\ 3,\ v_1v_2)$ and the sloop product yields an $S(\lambda_1\lambda_2;\ 2,\ 3,\ v_1v_2 + v_1+v_2)$. An $S(\lambda;\ 3,\ 4,\ v)$ is equivalent with an $ES(\lambda;\ 3,\ 4,\ v)\nu$ satisfying $\nu([x,x,y,y]) = \lambda$ for all $x,\ y \in S$. This identification yields a direct product for $S(\lambda;\ 3,\ 4,\ v)$. Given two $S(\lambda_i;\ 3,\ 4,\ v_i)$ on sets S_i, $i = 1,\ 2$, this product produces an $S(\lambda_1\lambda_2;\ 3,\ 4,\ v_1v_2)$ on $S_1 \times S_2$. More sophisticated ways of obtaining product constructions for $S(\lambda;\ 2,k,\ v)$ by associating algebraic structures with them are reviewed in [5]. In particular, in [5] a construction yielding an $S(2,\ k,\ v_1v_2)$ from an $S(2,\ k,\ v_1)$ and an $S(2,\ k,\ v_2)$ is given for $k = p^n$, p a prime. Another construction described in [5] yields an $S(2,\ k, (k-2)v_1v_2+v_1+v_2)$ from an $S(2,\ k,\ v_1)$ and an $S(2,\ k,\ v_2)$ for $k = p^n+1$, p a prime. These two constructions generalize the two product constructions described above for $k = 3$. However, for $k > 3$, the constructed product design is not uniquely determined by the two component designs, but depends on some additional algebraic structure put on these two designs. Of all the product constructions for t-designs described above and in [5], the only one that works for $t > 2$ is the direct product for $S(\lambda;\ 3,\ 4,\ v)$. None of them works for large sets.

5. REPETITION-TRIVIAL ARRAYS AND t-TRIVIAL DESIGNS.

In this section, we will study a product construction for a class of t-designs, namely t-trivial $S(\lambda;\ t,\ t+1,\ v)$. (These will be defined somewhat later on.) Unlike the product constructions for t-designs described in section 4, this one will work for all t and for large sets. The importance of t-trivial $S(\lambda;\ t,\ t+1,\ v)$, and equivalent structures, lies in their use in the construction of non-trivial simple t-designs for all t [19, 21].

If $f:S_1 \to S_2$ is a mapping and B is an S_1-multiset, then $f(B)$ denotes the S_2-multiset obtained by replacing every element b of B by $f(b)$. Thus, we can extend every mapping f from S_1 to S_2 to a mapping from $M(S_1)$ to $M(S_2)$. In turn, we can extend this mapping to a mapping from $M(M(S_1))$ to $M(M(S_2))$ and so on. Thus, if μ is an $M(S_1)$-multiset, then $f(\mu)$ is an $M(S_2)$-multiset. (The preceding conventions extend standard conventions for ordinary sets to multisets.) If μ_1 is an $M(S_1)$-multiset and μ_2 is an $M(S_2)$-multiset, then an <u>isomorphism</u> between μ_1 and μ_2 is a bijection $f:S_1 \to S_2$ such that $f(\mu_1) = \mu_2$. An <u>automorphism</u> of an $M(S)$-multiset μ is an isomorphism between μ and μ. We denote the automorphism group of μ by $\mathrm{Aut}(\mu)$. If $Y \subset S$, we call μ <u>Y-trivial</u> if $\mathcal{S}_Y \subset \mathrm{Aut}(\mu)$, where we identify $\sigma \in \mathcal{S}_Y$ with the permutation σ_0 of S with $\sigma_0|Y = \sigma$ and $\sigma_0|(S-Y) = \mathrm{id}_{S-Y}$. We call μ <u>t-trivial</u> if μ is Y-trivial for at least one t-subset Y of S.

We call a k-S-array μ <u>repetition-trivial</u> if $\mu(x_1, \ldots, x_k) = \mu(y_1, \ldots, y_k)$ whenever $\{x_1, \ldots, x_k\} = \{y_1, \ldots, y_k\}$. Similarly, we call an $M_k(S)$-multiset μ <u>repetition-trivial</u> if $\mu([x_1, \ldots, x_k]) = \mu([y_1, \ldots, y_k])$ whenever $\{x_1, \ldots, x_k\} = \{y_1, \ldots, y_k\}$. Clearly, every repetition-trivial array is totally symmetric. The canonical one-to-one correspondence between totally symmetric k-S-arrays and $M_k(S)$-multisets induces a one-to-one correspondence between repetition-trivial k-S-arrays and repetition-trivial $M_k(S)$-multisets. Obviously, the direct product of two repetition-trivial k-S_i-arrays ($M_k(S_i)$-multisets), $i=1, 2$, is repetition-trivial.

In the remainder of this section, S will be a v-set, $v \neq 0$, and $\{\infty_1, \ldots, \infty_t\}$ will be a t-set disjoint from S. If ε is a $P_{1,t+1}(S)$-multiset, then we can define a repetition-trivial $M_{t+1}(S)$-multiset μ_ε by putting, for each $B \in M_{t+1}(S)$, $\mu_\varepsilon(B) = \varepsilon(s(B))$. Similarly, we can define an $\{\infty_1, \ldots, \infty_t\}$-trivial $P_{t+1}(S \cup \{\infty_1, \ldots, \infty_t\})$-multiset ν_ε by putting, for each $B \in P_{t+1}(S \cup \{\infty_1, \ldots, \infty_t\})$, $\nu_\varepsilon(B) = \varepsilon(B \cap S)$. It is easy to see that the mappings $\varepsilon \to \mu_\varepsilon$ and $\varepsilon \to \nu_\varepsilon$ define one-to-one correspondences between the set of all (simple) $P_{1,t+1}(S)$-multisets, the set of all (simple) repetition-trivial $M_{t+1}(S)$-multisets and the set of all (simple)

$\{\infty_1, \ldots, \infty_t\}$-trivial $P_{t+1}(S \cup \{\infty_1, \ldots, \infty_t\})$-multisets. It is easy to see that if $A \in P_{0,t}(S)$ and if $B \in P_{t-|A|}(\{\infty_1, \ldots, \infty_t\})$, then the number of blocks of ν_ε containing $A \cup B$ equals

$$|A|\,\varepsilon(A) + \sum_{x \in S-A} \varepsilon(A \cup \{x\}) = \sum_{x \in S} \varepsilon(A \cup \{x\}).$$

Similarly, if $C \in M_n(S)$, $0 \leq n \leq t$, and $s(C) = A$, then the number of blocks of μ_ε of the type $C + (t+1-n)\{x\}$, where x runs over all elements of S, also equals $\sum_{x \in S} \varepsilon(A \cup \{x\})$. This means that μ_ε is $(\lambda; n)$-regular, $1 \leq n \leq t$, iff $\sum_{x \in S} \varepsilon(A \cup \{x\}) = \lambda$ for all $A \in P_{1,n}(S)$ iff every t-subset B of $S \cup \{\infty_1, \ldots, \infty_t\}$ with $1 \leq |B \cap S| \leq n$ is contained in exactly λ blocks of ν_ε. Thus, every $(\lambda; n)$-regular repetition-trivial $M_{t+1}(S)$-multiset is also $(\lambda; n')$-regular for all n' with $1 \leq n' \leq n$. In particular, every repetition-trivial $ES(\lambda; t, t+1, v)$ is $(\lambda; n)$-regular for all n with $1 \leq n \leq t$. However, if μ is an idempotent $ES(\lambda; 2, 3, v)$, $v > 1$, then μ is repetition-trivial, but μ is $(\lambda v; 0)$-regular instead of $(\lambda; 0)$-regular. Note that, for any $P_{1,t+1}(S)$-multiset ε, μ_ε is $(\sum_{x \in S} \varepsilon(\{x\}); 0)$-regular and $\{\infty_1, \ldots, \infty_t\}$ is contained in exactly $\sum_{x \in S} \varepsilon(\{x\})$ blocks of ν_ε. A $P_{1,t+1}(S)$-multiset ε is called an $\aleph_t$-$S(\lambda; t, t+1, v+t)$ if $\sum_{x \in S} \varepsilon(A \cup \{x\}) = \lambda$ for all $A \in P_{0,t}(S)$. (Note that an $\aleph_t - S(\lambda; t, t+1, v+t)$ has v points and not $v+t$. Our notation is motivated by the very close relationship between $\aleph_t - S(\lambda; t, t+1, v+t)$ and t-trivial $S(\lambda; t, t+1, v+t)$.) It follows from the above that a $P_{1,t+1}(S)$-multiset ε is an $\aleph_t - S(\lambda; t, t+1, v+t)$ iff ν_ε is an $S(\lambda; t, t+1, v+t)$ iff μ_ε is a regular $ES(\lambda; t, t+1, v)$. Thus, there are one-to-one correspondences between the set of all (simple) $\{\infty_1, \ldots, \infty_t\}$-trivial $S(\lambda; t, t+1, v+t)$ on $S \cup \{\infty_1, \ldots, \infty_t\}$, the set of all (simple) $\aleph_t - S(\lambda; t, t+1, v+t)$ on S, the set of all (simple) repetition-trivial regular $ES(\lambda; t, t+1, v)$ on S and the set of all (simple) repetition-trivial regular $OA(\lambda; t, t+1, v)$ on S. As the direct product of two repetition-trivial regular $OA(\lambda_i; t, t+1, v_i)$, $i = 1, 2$, is a repetition-trivial regular $OA(\lambda_1\lambda_2; t, t+1, v_1v_2)$, this means that we have implicitly defined a product construction for t-trivial $S(\lambda; t, t+1, v)$, yielding a (simple) t-trivial $S(\lambda_1\lambda_2; t, t+1, v_1v_2 + t)$ from (simple) t-trivial $S(\lambda_i; t, t+1, v_i + t)$, $i = 1, 2$. We now will give a direct description of this product in terms of $\mathcal{Y}_t - S(\lambda;\ t,\ t+1,\ v+t)$. Let S_i be a v_i-set, $i = 1, 2$. We define $pr_i : S_1 \times S_2 \to S_i$ by $pr_i(x_1, x_2) = x_i$, $i = 1, 2$. If ε_i is an $\mathcal{Y}_t - S(\lambda_i;\ t,\ t+1,\ v_i+t)$ on S_i, $i = 1, 2$, then the unique $\mathcal{Y}_t\text{-}S(\lambda_1\lambda_2;\ t,\ t+1,\ v_1v_2 + t)$ $\varepsilon_1 \times \varepsilon_2$ on $S_1 \times S_2$ with $\mu_{\varepsilon_1 \times \varepsilon_2} = \mu_{\varepsilon_1} \times \mu_{\varepsilon_2}$ can be defined directly from ε_1 and ε_2 by putting $(\varepsilon_1 \times \varepsilon_2)(A) = \varepsilon_1(pr_1 A) \cdot \varepsilon_2(pr_2 A)$ for all $A \in P_{1,t+1}(S_1 \times S_2)$. Retranslating everything in terms of t-trivial $S(\lambda;\ t,\ t+1,\ v+t)$ is easy, but the construction seems more natural in terms of $\mathcal{Y}_t\text{-}S(\lambda;\ t,\ t+1,\ v+t)$. In general, even though $\mathcal{Y}_t - S(\lambda;\ t,\ t+1,\ v+t)$ are the least familiar of the four equivalent structures, they nevertheless turn out, in practice, to be the easiest to work with.

A large set of disjoint $\mathcal{Y}_t - S(\lambda;\ t,\ t+1,\ v+t)$, briefly $\mathcal{Y}_t\text{-}LS(\lambda;\ t,\ t+1,\ v+t)$ on S is a collection $(\varepsilon_r)_{r\in R}$ of $\mathcal{Y}_t$-$S(\lambda;\ t,\ t+1,\ v+t)$ on S such that $\sum_{r\in R} \varepsilon_r = P_{1,t+1}(S)$. An $LS(\lambda;\ t,\ t+1,\ v)$ is called Y-trivial if all its members are Y-trivial. An $LES(\lambda;\ t,\ t+1,\ v)$ or $LA(\lambda;\ t,\ t+1,\ v)$ is called repetition-trivial if all its members are repetition-trivial. It is easy to check that if $(\varepsilon_r)_{r\in R}$ is a collection of $\mathcal{Y}_t\text{-}S(\lambda;\ t,\ t+1,\ v+t)$ on S, then $(\varepsilon_r)_{r\in R}$ is an $\mathcal{Y}_t\text{-}LS(\lambda;\ t,\ t+1,\ v+t)$ iff $(\nu_{\varepsilon_r})_{r\in R}$ is an $LS(\lambda;\ t,\ t+1,\ v+t)$ iff $(\mu_{\varepsilon_r})_{r\in R}$ is an $LES(\lambda;\ t,\ t+1,\ v)$. It follows that there are one-to-one correspondences between the set of all $\mathcal{Y}_t\text{-}LS(\lambda;\ t,\ t+1,\ v+t)$ on S, the set of all $\{\infty_1, \ldots, \infty_t\}$-trivial $LS(\lambda;\ t,\ t+1,\ v+t)$ on $S \cup \{\infty_1, \ldots, \infty_t\}$, the set of all repetition-trivial regular $LES(\lambda;\ t,\ t+1,\ v)$ on S and the set of all repetition-trivial regular $LA(\lambda;\ t,\ t+1,\ v)$ on S. As a consequence, the product construction for t-trivial $S(\lambda;\ t,\ t+1,\ v+t)$ produces large sets from large sets.

The main difficulty in using the above product construction for t-trivial $S(\lambda;\ t,\ t+1,\ v)$, to construct non-trivial simple $S(\lambda;\ t,\ t+1,\ v)$ for all t, is that it only produces a non-trivial simple t-design if we start with two simple t-trivial t-designs, at least one of which is non-trivial. Most natural direct product constructions for any kind of structure suffer from the defect that if any of the two components has repeated blocks (rows), then so does the direct product. However, nearly all natural direct product constructions can be generalized considerably and adapted in such a way that they allow, at least in certain instances, to get rid of repeated blocks. In particular, this is the case for the product construction for t-trivial S(λ; t, t+1, v). The details are rather technical. They are given in [21, Proposition 6] in terms of $\mathcal{X}_t$ - S(λ; t, t+1, v). We will not reproduce them here. The resulting construction, namely Proposition 6 of [21], is the cornerstone of the inductive proof of the existence of nontrivial simple t-designs for all t, given in section 2 of [21]. (This proof is a simplified version of the original proof given in [19], which used a more powerful and more sophisticated construction. We will mention this construction in the next section, as Proposition 2.)

6. OTHER APPLICATIONS OF REGULAR ES(λ; t, t+1, v).

A recurrent construction for LS(2, 3, v) found independently, in different versions, by Schreiber [15], Lu [8] and the author [17], allows to construct an LS(2, 3, vw+2) from an LS(2, 3, w+2), an LS(2, 3, v+2)$(\nu_r)_{r\in R}$ on $S \cup \{\infty_1, \infty_2\}$, S a v-set, and an LES(2, 3, v) $(\mu_r)_{r\in R}$ on S such that $\mu_r^* = \nu_r \| S$ for all r∈R. Neither [8] nor [17] state the construction in these terms, but it is easy to either retranslate Lu's result or generalize the author's result in [17] (see also [12]) to obtain this theorem. The research that eventually led to the original proof of the existence of non-trivial simple t-designs for all t, given in [19], actually started out with a generalization of this construction, in an ordered version involving OD(λ; t, t+1, v), to arbitrary t and λ. This generalization allows, for instance, to produce LS(t, t+1, v^n+t) for all n ∈ N, given an LS(t, t+1, v+t) with additional structure. Unfortunately, the required additional structure is rather awkward and not very natural. For t ≥ 3, this structure is a lot more restrictive and more complicated than a pair $((\nu_r)_{r\in R}, (\mu_r)_{r\in R})$, where $(\nu_r)_{r\in R}$ is an LS(t, t+1, v+t) on $S \cup \{\infty_1, \ldots, \infty_t\}$, S a v-set, and $(\mu_r)_{r\in R}$ is an LES(t, t+1, v) on S such that $\mu_r^* = \nu_r \| S$ for all r∈R. The required structure easily allows to construct such a pair, but, for t ≥ 3, the converse is not true. (Because the extra structure is so complicated and because, even for $\lambda > 1$, we do not know any concrete applications with $t \geq 3$ not contained in Proposition 2 below, we do not give the details of the construction here. Essentially, they consist in applying Propositions 2.3 and 3.3 of [19] in a direction suggested by Proposition 3.4 of [19].) When one generalizes a natural theorem and obtains an extremely technical theorem, which is difficult to manipulate, it very often pays off to generalize this technical theorem even further and then hunt for simplifications in directions other than the one we started from. In this case, this led to the following proposition, which was the implicit cornerstone of the proof of the existence of non-trivial simple t-designs for arbitrary t, given in [19], even though this proposition was not explicitly mentioned in [19]. (It is a combination of Propositions 2.3, 3.3, 3.4 and 3.5 of [19].) An explicit proof is given in [21, Propositions 8 and 12].

<u>Proposition 2</u>: Let $(\varepsilon_r)_{r\varepsilon R}$ be a collection of $\mathscr{S}_t$-$S(\lambda;\ t,\ t+1,\ u+t)$ on a u-set X such that $\sum_{r\varepsilon R} \varepsilon_r = w \cdot P_{1,t+1}(X)$, $w \geq 1$. Assume that, for each $B \varepsilon P_{1,t+1}(X)$, a positive integer $\lambda(B)$ is given such that $\lambda(B)$ divides $\varepsilon_r(B)$ for all $r\varepsilon R$. Assume that, for each $B \varepsilon P_{1,t+1}(X)$, there is an $LS(\lambda(B);\ t+1-|B|,\ t+2-|B|,\ w+t+1-|B|)$. Then there is an $LS(\lambda;\ t,\ t+1,\ uw+t)$.

The proof of Proposition 2 uses the fact that the required extra structure, mentioned above, can always be put on a t-trivial $S(\lambda;\ t,\ t+1,\ v+t)$. In [19], this is done explicitly in Proposition 3.5. In [21] it is done implicitly, but in a more intuitive way than in [19], in section 4. However, t-trivial $S(\lambda;\ t,\ t+1,\ v+t)$ are the only t-designs with $t \geq 3$ for which we know how to handle the extra structure.

If ν is an $S(\lambda;\ t,\ k,\ v)$ on a set S and if G is any subgroup of $\mathscr{S}_S$, then obviously, $\mu = \sum_{f\varepsilon G} f(\nu)$ is an $S(\lambda|G|;\ t,\ k,\ v)$ with $G \subset \mathrm{Aut}(\mu)$. In fact, one can divide, in this $S(\lambda|G|;\ t,\ k,\ v)$, the multiplicities of all blocks by the greatest common divisor λ_0 of all these multiplicities, and thus, obtain an $S(\lambda|G|/\lambda_0;\ t,\ k,\ v)$. In particular, if ν is an $S(\lambda;\ t,\ k,\ v)$ on a v-set S and if $Y \subset S$, then one can always construct a Y-trivial $S(\lambda';\ t,\ k,\ v)$ on S for some λ' dividing $|Y|!\lambda$. Actually, for Y-trivial $S(\lambda; t, k, v)$ one can adapt the procedure in such a way that one does not need an $S(\lambda; t, k, v)$ on S to start with, but only a structure on $S-Y$ that looks combinatorially like an $S(\lambda; t, k, v)$ with $|Y|$ points deleted. Such a structure may exist, even when an $S(\lambda; t, k, v)$ does not. In particular, the condition that λ should be divisible by $\lambda(t, t+1, v+t) = \gcd(v, \ell cm\{1, \ldots, t+1\})$ is necessary for the existence of an $S(\lambda; t, t+1, v+t)$, but not sufficient. By Proposition 1 (2), it is sufficient for the existence of a regular $ES(\lambda; t, t+1, v)$. Regular $ES(\lambda; t, t+1, v)$ look combinatorially very much like an $S(\lambda; t, t+1, v+t)$ with t points deleted and thus, can be used to construct t-trivial $S(\lambda'; t; t+1, v+t)$. The details are given in [21]. All applications of Proposition 2 given in [21] use collections $(\varepsilon_r)_{r\in R}$ of $\mathcal{X}_t$-$S(\lambda; t, t+1, u+t)$ constructed using the preceding ideas, in some cases combined with some very simple additional observations. One possible way to obtain smaller values of λ than those obtained in [21] would be to try to find collections $(\varepsilon_r)_{r\in R}$ for which Proposition 2 would yield better results.

Proposition 2 looks rather silly for $t=2$ and $t=3$, because much better is available in these cases. These better techniques are contained in results in [14, 16, 18]. In [18], regularity properties of arrays are used explicitly. Because regular arrays are not used explicitly in [14, 16], we state the following proposition, which is a generalization of constructions used in the proofs of Proposition 1 of [14] and Lemma 1 of [16]. Checking the proposition is straightforward.

<u>Proposition 3</u>: Let $(\mu_r)_{r\in R}$ be a regular $LES(\lambda; 2, 3, v)$ on a v-set S. Let $\{\infty_1, \infty_2\}$ be a 2-set with $\{\infty_1, \infty_2\} \cap (S \times Z_2) = \emptyset$. Put an arbitrary tournament T on S. Then we can construct an $LS(2\lambda; 2, 3, 2v+2)(\nu_r)_{r\in R}$ on $\{\infty_1, \infty_2\} \cup (S \times Z_2)$, where ν_r is obtained from μ_r by replacing:

(a) each block $[x, y, z]$ of μ_r with $|\{x, y, z\}| = 3$ by the eight blocks $\{(x, i), (y, j), (z, k)\}$, $\{i, j, k\} \subset Z_2$.

(b) each block $[x, x, y]$ of μ_r with $x \neq y$, by the six blocks $\{(x, 0), (x, 1), (y, 0)\}$, $\{(x, 0), (x, 1), (y, 1)\}$, $\{(x, 0), (y, 0), \infty_i\}$, $\{(x, 1), (y, 1), \infty_i\}$, $\{(x, 0), (y, 1), \infty_j\}$, and $\{(x, 1), (y, 0), \infty_j\}$, where $i=1$ and $j=2$ if $(x, y) \in T$ and $i=2$

and $j=1$ if $(x, y) \notin T$.
(c) each block $[x, x, x]$ of μ_r by the four 3-subsets of $\{\infty_1, \infty_2, (x, 0), (x, 1)\}$.

Another construction in [16] allows to construct an LS(λ; 2, 3, 2v-2) from an LS(λ; 2, 3, v), for all even λ. This construction does not use regular arrays, but is nevertheless related in spirit to the other constructions in [14, 16, 18]. One of the main differences between the proof of Proposition 2 and the more efficient techniques for $t \in \{2, 3\}$ is that, in the proof of Proposition 2, designs are built up by using exclusively certain structures, called S(λ; t, t+1, δ) in [21], as building blocks. The more efficient techniques use ad hoc building blocks. Possibly, the correct generalization of these ad hoc structures, and the corresponding constructions, to higher t may yield non-trivial simple t-designs for arbitrary t and relatively small λ. (Actually, finding some "natural" generalizations of, for instance, the techniques of Proposition 3, to higher t is relatively easy. However, the required building blocks seem difficult to construct for $t \geq 4$. The hard problem is, of course, to generalize the techniques in such a way that they produce concrete results.)

In most of this paper, we have restricted our attention to the case $k = t+1$. Let us close by making some remarks about arbitrary k. It is well known and easy to see that every (L)S(λ; k-1, k, v) is also an (L)S($\lambda\binom{v-t}{k-1-t}/(k-t)$; t, k, v) for all $t \leq k-1$. Thus, in a sense, the techniques in [19, 21] actually work for arbitrary k, because they enable us to construct LS(λ; k-1, k, v). In trying to find less trivial and more efficient applications of our techniques for $k > t + 1$, we encounter several difficulties. Some of these will be discussed in [22], in the more general context of designs on posets. In [22], we will also discuss some potential ways of getting around the difficulties, but for the moment, we do not have concrete applications.

ACKNOWLEDGMENTS

Research supported by NSF grant DMS-8815112 and NSA grant MDA 904-88-H-2005.

REFERENCES

[1] F. E. Bennett and E. Mendelsohn, Extended (2,4)-designs, J. Combinatorial Theory A, 29 (1980), 74-86.

[2] T. Beth, On resolutions of Steiner systems, Mitteilungen aus dem Mathem. Seminar Giessen, Heft 136 (1979).

[3] T. Beth, D. Jungnickel and H. Lenz, Design Theory, Bibliographisches Institut, Mannheim-Wien-Zurich, 1985.

[4] E. J. Billington, Designs with repeated elements in blocks: a survey and some recent results, to appear.

[5] B. Ganter and H. Werner, Co-ordinatizing Steiner systems, Annals of Discrete Math. 7 (1980), 3-24.

[6] J. E. Graver and W. B. Jurkat, The module structure of integral designs, J. Combinatorial Theory, Ser. A 15 (1973), 75-90.

[7] D. M. Johnson and N. S. Mendelsohn, Extended triple systems, Aequationes Math. 8 (1972), 291-298.

[8] J. X. Lu, On large sets of disjoint Steiner triple systems I, J. Combinatorial Theory, Ser. A 34 (1983), 140-146.

[9] S. S. Magliveras and D. W. Leavitt, Simple six designs exist, Congressus Numerantium, Vol 40 (1983), 195-205.

[10] E. H. Moore, Concerning triple systems, Math. Ann. 43 (1893), 271-285.

[11] D. Raghavarao, Construction and combinatorial problems in design of experiments, Wiley, New York, 1971.

[12] A. Rosa, Intersection properties of Steiner systems, Annals of Discrete Math. 7 (1980), 115-128.

[13] S. Schreiber, Covering all triples on n marks by disjoint Steiner systems, J. Combinatorial Theory (A) 15 (1973), 347-350.

[14] S. Schreiber, Some balanced complete block designs, Israel J. Math. 18 (1974), 31-37.

[15] S. Schreiber, Personal communication (1977).

[16] L. Teirlinck, On the maximum number of disjoint triple systems, J. Geometry 6 (1975), 93-96.

[17] L. Teirlinck, Combinatorial Structures, Thesis, Vrije Universiteit Brussel, 1976.

[18] L. Teirlinck, On large sets of disjoint quadruple system, Ars Combinatoria, Vol. 17 (1984), 173-176.

[19] L. Teirlinck, Non-trivial t-designs without repeated blocks exist for all t, Discrete Math. 65 (1987), 310-311.

[20] L. Teirlinck, On large sets of disjoint ordered designs, Ars Combinatoria, Vol. 25 (1988), 31-37.

[21] L. Teirlinck, Locally trivial t-designs and t-designs without repeated blocks, to appear.

[22] L. Teirlinck, Designs on products of posets, in preparation.

[23] R. M. Wilson, The necessary conditions for t-designs are sufficient for something, Utilitas Math. 4 (1973), 207-215.

[24] R. M. Wilson, Some partitions of all triples into Steiner triple systems, Hypergraph Seminar, Ohio State Univ. 1972. Lecture Notes Math. 411 (1974), 267-277, Springer, Berlin.

Department of Algebra, Combinatorics and Analysis
Auburn University, Alabama 36849 U.S.A.

The 'Snake Oil' method for proving combinatorial identities

Herbert S. Wilf*

I have recently been working on a book about generating functions. It will be called 'Generatingfunctionology,' and it is intended to be an upper-level undergraduate, or graduate text in the subject. The object is to try to impress students with the beauty of this subject too, so they won't think that only bijections can be lovely.

In one section of the book I will discuss combinatorial identities, and the approach will be this. First I'll give the 'Snake Oil' method, in the spirit of a unified approach that works on many relatively simple identities. It involves generating functions. Second, I will write about a much more powerful method that works on 'nearly all' identities, including all classical hypergeometric identities and many, many binomial coefficient identities.

These two approaches will be mirrored here, in that this article will mostly be about the Snake Oil method, whereas the talk that I will give at the conference will be about the much more powerful method of WZ pairs [WZ], which at this writing is still under development. A brief summary of the WZ results appears in section (II) below.

Aside from these developments, there have been other unifying forces at work in the field of identities. The expository paper of Roy [Ro], shows how even without a computer one can recognize many binomial identities as cases of just a very few identities in the theory of hypergeometric series. The work of Knuth [Kn] shows how a few rules about binomial coefficients and their handling can, in skilled hands, prove many difficult identities. This point is reinforced in [GKP].

A third method of great generality is due to Egorychev [Eg], and is called by him the 'method of residues.' It is quite strong on binomial coefficient identities, but does not have the scope of the WZ machine.

(I) The 'Snake Oil' method

The first method that I want to discuss here is the 'Snake Oil' method, after a slang expression for a quack medicine or nostrum,† that is highly touted as a panacæa, but in fact falls quite short. I hope you won't think it falls quite short, because I'm not touting it as a cure-all. It is, though, an excellent medicine to try

* Research supported in part by the United States Office of Naval Research

† The Random House Dictionary of the English Language states that 'snake oil' is of American origin. ca. 1925-30, and that a typical use of the expression is *The governor promised to lower taxes, but it was the same old snake oil.*

for whatever sum ails you.

The ingredients of the method are the following.

(a) One needs to observe the convention that the binomial coefficient $\binom{n}{k}$ vanishes if $k < 0$, or if n and k are both integers and $0 \le n < k$.

(b) One needs to observe the convention that if a certain summation extends over an index k, say, and if the range of summation is not specified, then that range is from $-\infty$ to ∞.

(c) One needs to believe the binomial theorem

$$\sum_k \binom{n}{k} x^k = (1+x)^n, \tag{1}$$

which is valid for all x if n is a nonnegative integer, and for $|x| < 1$ otherwise.

(d) One needs the series expansion

$$\sum_{k \ge 0} \binom{k}{b} x^k = \frac{x^b}{(1-x)^{b+1}} \tag{2}$$

which is valid for nonnegative integer b and $|x| < 1$.

The basic idea of the method is what I might call the *external* approach to identities rather than the usual *internal* method.

To explain the difference between these two points of view, suppose we want to prove some identity that involves binomial coefficients. Typically such a thing would assert that some fairly intimidating-looking sum is in fact equal to such-and-such a simple function of n (an excellent collection of these is Gould [Gou]).

One approach that is now customary consists primarily in looking inside the summation sign ('internally'), and using binomial coefficient identities or other manipulations of indices *inside* the summations, to bring the sum to manageable form.

In the *external*, or generatingfunctionological, approach, one begins by giving just a quick glance at the expression that is inside the summation sign, long enough to spot the 'free variables,' i.e., what it is that the sum depends on after the dummy variables have been summed over. Suppose that such a free variable is called n.

Then instead of trying to grapple with the sum, the trick is to find the generating function whose coefficients are the values of your sum. More precisely, if $f(n)$ is your sum, instead of going *inside* to evaluate $f(n)$, go *outside* to evaluate $\sum_n f(n)x^n$. Then after you've done that, read off the coefficient of x^n, and you're finished. Here it is, a little more systematically:

(a) Identify the free variable, say n, that the sum depends on. Give a name to the sum that you are working on: call it $f(n)$.

(b) Let $F(x)$ be the ordinary power series generating function (opsgf) whose coefficient of x^n is $f(n)$, the sum that you'd like to evaluate.

(c) Multiply the sum by x^n, and sum on n. Your generating function is now expressed as a double sum, over n, and over whatever variable was first used as a dummy summation variable.

(d) Interchange the order of the two summations that you are now looking at, and perform the inner one in simple closed form. For this purpose it will be helpful to have a catalogue of series whose sums are known.
(e) Try to identify the coefficents of the generating function of the answer, because those coefficients are what you want to find.

Several examples

1. A Fibonacci sum

Consider the sum

$$\sum_{k\geq 0}\binom{k}{n-k} \qquad (n=0,1,2,\ldots).$$

The free variable is n, so let's call the sum $f(n)$. Write it out like this:

$$f(n)=\sum_{k\geq 0}\binom{k}{n-k}.$$

Multiply both sides by x^n and sum over $n \geq 0$. You have now arrived at step (c) of the general method, and you are looking at

$$F(x)=\sum_{n\geq 0}x^n\sum_{k\geq 0}\binom{k}{n-k}.$$

Ready for step (d)? Interchange the sums, to get

$$F(x)=\sum_{k\geq 0}\sum_{n\geq 0}\binom{k}{n-k}x^n.$$

We would like to 'do' the inner sum, the one over n. The trick is to get the exponent of x to be exactly the same as the index that appears in the binomial coefficient. In this example the exponent of x is n, and n is involved in the downstairs part of the binomial coefficient in the form $n-k$. To make those the same, the correct medicine is to multiply inside the sum by x^{-k} and outside the inner sum by x^k, to compensate. The result is

$$F(x)=\sum_{k\geq 0}x^k\sum_{n\geq 0}\binom{k}{n-k}x^{n-k}.$$

Now the exponent of x is the same as the lower index of the binomial coefficient. Hence take $r=n-k$ as the new dummy variable of summation in the inner sum. We find then

$$F(x)=\sum_{k\geq 0}x^k\sum_{r}\binom{k}{r}x^r.$$

We recognize the inner sum immediately, as $(1+x)^k$. Hence

$$F(x) = \sum_{k\geq 0} x^k(1+x)^k = \sum_{k\geq 0}(x+x^2)^k = \frac{1}{1-x-x^2}.$$

The generating function on the right is an old friend; it generates the Fibonacci numbers (if you didn't recognize that, a partial fraction expansion would produce the coefficients quite explicitly). Hence $f(n) = F_n$, and we have discovered that

$$\sum_{k\geq 0}\binom{k}{n-k} = F_n \qquad (n = 0, 1, 2, \ldots).$$

2. A bit harder

Consider the sum

$$\sum_{k\geq 0}\binom{n+k}{m+2k}\binom{2k}{k}\frac{(-1)^k}{k+1} \qquad (m, n \geq 0). \tag{3}$$

Let $f(n)$ denote the sum in question (m would work just as well), and let $F(x)$ be its opsgf. Dive in immediately by multiplying by x^n and summing over $n \geq 0$, to get

$$\begin{aligned}
F(x) &= \sum_{n\geq 0} x^n \sum_{k\geq 0}\binom{n+k}{m+2k}\binom{2k}{k}\frac{(-1)^k}{k+1}\\
&= \sum_{k\geq 0}\binom{2k}{k}\frac{(-1)^k}{k+1}x^{-k}\sum_{n\geq 0}\binom{n+k}{m+2k}x^{n+k}\\
&= \sum_{k\geq 0}\binom{2k}{k}\frac{(-1)^k}{k+1}x^{-k}\frac{x^{m+2k}}{(1-x)^{m+2k+1}} \qquad \text{(by (2))}\\
&= \frac{x^m}{(1-x)^{m+1}}\sum_{k\geq 0}\binom{2k}{k}\frac{1}{k+1}\left\{\frac{-x}{(1-x)^2}\right\}^k\\
&= \frac{-x^{m-1}}{2(1-x)^{m-1}}\left\{1-\sqrt{1+\frac{4x}{(1-x)^2}}\right\}\\
&= \frac{-x^{m-1}}{2(1-x)^{m-1}}\left\{1-\frac{1+x}{1-x}\right\}\\
&= \frac{x^m}{(1-x)^m}.
\end{aligned}$$

The original sum is therefore the coefficient of x^n in the last member above. But that is $\binom{n-1}{m-1}$, by (2) again, and we have our answer.

If the manipulations seemed long, consider that at least they're always the *same* manipulations, whenever the method is used, and also that with some effort a computer could be trained to do them!

3. An equality

Suppose we have two complicated sums and we want to show that they're the same. Then the generating function method, if it works, should be very easy to carry out. Indeed, one might just find the generating functions of each of the two sums independently and observe that they are the same.

Suppose we want to prove that

$$\sum_k \binom{m}{k}\binom{n+k}{m} = \sum_k \binom{m}{k}\binom{n}{k}2^k \qquad (n, m \geq 0),$$

without evaluating either of them.

Multiply on the left by x^n, sum on $n \geq 0$ and interchange the summations, to arrive at

$$\begin{aligned}\sum_k \binom{m}{k} x^{-k} \sum_{n\geq 0} \binom{n+k}{m} x^{n+k} &= \sum_k \binom{m}{k} x^{-k} \frac{x^m}{(1-x)^{m+1}} \\ &= \frac{x^m}{(1-x)^{m+1}}\left(1+\frac{1}{x}\right)^m \\ &= \frac{(1+x)^m}{(1-x)^{m+1}}.\end{aligned}$$

If we multiply on the right by x^n, etc., we find

$$\begin{aligned}\sum_k \binom{m}{k} 2^k \sum_{n\geq 0} \binom{n}{k} x^n &= \frac{1}{(1-x)} \sum_k \binom{m}{k}\left(\frac{2x}{(1-x)}\right)^k \\ &= \frac{1}{(1-x)}\left(1+\frac{2x}{1-x}\right)^m \\ &= \frac{(1+x)^m}{(1-x)^{m+1}}.\end{aligned}$$

Since the generating functions are the same, the sums are equal, and we're finished.

4. An identity of D. E. Knuth

Here we want to evaluate, for $m, n \geq 0$,

$$\sum_k \binom{n}{k}\binom{n-k}{\lfloor \frac{m-k}{2} \rfloor} y^k. \tag{4}$$

The appearance of the 'floor' function is a bit disquieting, and in fact, it removes this sum from the realm of identities of hypergeometric type. Nevertheless, Snake Oil is more than equal to the occasion.

We are going to multiply the sum by x-to-the-something, and sum over the 'something'. There are two choices: m, or n, because there are two free variables. The one to use is m, because it appears in only one place inside the sum, hence after the step of interchanging orders of summation, that sum will have only one binomial coefficient in it.

Call the unknown sum $f_m(y)$. Multiply (4) by x^m and sum on $m \geq 0$, to get

$$F(x;y) = \sum_k \binom{n}{k} y^k \sum_{m\geq 0} \binom{n-k}{\lfloor \frac{m-k}{2} \rfloor} x^m.$$

The next step is that inside the sum we make the exponent of x into $m-k$, to obtain

$$\begin{aligned} F(x;y) &= \sum_k \binom{n}{k} (xy)^k \sum_{m\geq 0} \binom{n-k}{\lfloor \frac{m-k}{2} \rfloor} x^{m-k} \\ &= \sum_k \binom{n}{k} (xy)^k \sum_{r\geq 0} \binom{n-k}{\lfloor r/2 \rfloor} x^r. \end{aligned}$$

This problem differs from the preceding ones in that we don't know offhand about the inner sum. But it's not hard to find that

$$\begin{aligned} \sum_{r\geq 0} \binom{q}{\lfloor r/2 \rfloor} x^r &= \binom{q}{0} + \binom{q}{0} x + \binom{q}{1} x^2 + \binom{q}{1} x^3 + \cdots \\ &= (1+x)(1+x^2)^q \qquad (q \geq 0). \end{aligned}$$

Putting it all together, our unknown sum (4) is the coefficient of x^m in

$$\begin{aligned} F(x;y) &= (1+x) \sum_k \binom{n}{k} (xy)^k (1+x^2)^{n-k} \\ &= (1+x)(1+xy+x^2)^n. \end{aligned}$$

That's as far as we can go without specifying y. If $y = \pm 2$ we can go further, to get

$$\sum_k \binom{n}{k} \binom{n-k}{\lfloor \frac{m-k}{2} \rfloor} 2^k = \binom{2n+1}{m}$$

and

$$\sum_k \binom{n}{k} \binom{n-k}{\lfloor \frac{m-k}{2} \rfloor} (-2)^k = (-1)^m \binom{2n}{m} \left\{ \frac{2n-2m+1}{2n-m+1} \right\}.$$

5. Not necessarily binomial coefficients

The Snake Oil method works not only on sums that involve binomial coefficients, but on all sorts of counting numbers, as this example shows.

Let $\{a_n\}$ and $\{b_n\}$ be two sequences whose exponential generating functions are respectively $A(x)$, $B(x)$. Suppose that the sequences are connected by

$$b_n = \sum_k s(n,k) a_k \qquad (n \geq 1), \tag{5}$$

where the s's are the Stirling numbers of the first kind. Let $A(x)$ and $B(x)$ denote the exponential generating functions of these sequences, and let's find out how $A(x)$ and $B(x)$ are related to each other, in view of (5).

To do that, multiply (5) by $x^n/n!$, sum over n and use the fact that the Stirling numbers satisfy

$$\sum_n \frac{s(n,k)}{n!} x^n = \frac{1}{k!}\left\{\log \frac{1}{1-x}\right\}^k \qquad (k \geq 0).$$

The result is that

$$\begin{aligned} B(x) &= \sum_n \frac{x^n}{n!} \sum_k s(n,k) a_k \\ &= \sum_{k \geq 0} a_k \left\{\log \frac{1}{1-x}\right\}^k \\ &= A\left(\log \frac{1}{1-x}\right). \end{aligned} \tag{6}$$

Hence the exponential generating function relation (6) is equivalent to the sequence relation (5). As an example of its use, take the $\{a_n\}$ to be the Bernoulli numbers $\{B_n\}$, so that $A(x) = x/(e^x - 1)$. Then (6) says that

$$B(x) = \frac{1-x}{x} \log \frac{1}{1-x} = -\sum_{n \geq 1} \frac{x^n}{n(n+1)},$$

which is clearly the exponential generating function of the sequence

$$\left\{-\frac{n!}{n(n+1)}\right\}_{n \geq 1}.$$

Consequently we have the identity

$$\sum_k s(n,k) B_k = -\frac{n!}{n(n+1)} \qquad (n \geq 1)$$

between the Bernoulli numbers and the Stirling numbers of the first kind.

6. A tough one

For our final example, we prove a difficult identity of Graham and Riordan, namely

$$f(m) = \sum_k \binom{2n+1}{2k}\binom{m+k}{2n} = \binom{2m+1}{2n} \qquad (m, n \geq 0). \tag{7}$$

First, notice that we singled out the variable m as the free variable, rather than n, which is also a free variable. We did that because the m appears in only one place on the left side of (7), while n appears in two places. That means that after we multiply by x^m, sum, and interchange the summations, the sum on m will have just a single binomial coefficient in it, rather than two.

This is a general characteristic of the Snake Oil method. It works best if there is a free variable that appears only once.

Next, we would normally multiply through in (7) by x^m and continue, but since the claimed answer has $2m+1$ in it, it might save time if we multiply by x^{2m+1} instead (it only saves time; the method works fine if we use x^m, but it's a little more complicated to use).

If we do that, the result is that the ordinary power series generating function $F(x)$ of the sums on the left side of (7) is

$$\begin{aligned}
\sum_{m\geq 0} f(m)x^{2m+1} &= \sum_{m\geq 0} x^{2m+1} \sum_k \binom{2n+1}{2k}\binom{m+k}{2n} \\
&= \sum_k \binom{2n+1}{2k} \sum_{m\geq 0} \binom{m+k}{2n} x^{2m+1} \\
&= \sum_k \binom{2n+1}{2k} x^{-2k+1} \sum_{m\geq 0} \binom{m+k}{2n} x^{2(m+k)} \\
&= \sum_k \binom{2n+1}{2k} x^{-2k+1} \sum_r \binom{r}{2n} x^{2r}.
\end{aligned}$$

The innermost sum is, by (2),

$$\frac{x^{4n}}{(1-x^2)^{2n+1}},$$

and therefore

$$\begin{aligned}
F(x) &= \frac{x^{4n+1}}{(1-x^2)^{2n+1}} \sum_k \binom{2n+1}{2k} x^{-2k} \\
&= \frac{x^{4n+1}}{(1-x^2)^{2n+1}} \left\{ \frac{1}{2}(1+x^{-1})^{2n+1} + \frac{1}{2}(1-x^{-1})^{2n+1} \right\} \\
&= \frac{x^{2n}}{2} \left\{ \frac{1}{(1-x)^{2n+1}} - \frac{1}{(1+x)^{2n+1}} \right\}.
\end{aligned}$$

We are now finished, since the coefficient of x^{2m+1} in the last member is clearly as shown on the right side of (7).

In general, whenever a free variable appears only once inside an unknown sum, Snake Oil may be the best medicine. The list of identities that can be handled by this simple, programmable device is very lengthy indeed.

(II) WZ Pairs

Here is a brief summary of the method of WZ pairs, which at the time of this writing is just being brought into final form.

The method provides machinery whereby a vast collection of combinatorial identities can be given one-line proofs. The proof certificate consists of a single rational function. Here is an example of a one-line proof of a famous and difficult identity.

Theorem (Dixon). *The identity*

$$\sum_k (-1)^k \binom{n+b}{n+k}\binom{n+c}{c+k}\binom{b+c}{b+k} = \frac{(n+b+c)!}{n!b!c!}$$

is true.

Proof: Take

$$R(n,k) = \frac{(c+1-k)(b+1-k)}{2(n+k)(n+b+c+1)}. \blacksquare$$

Here is a one-line proof of the famous hypergeometric identity of Saalschütz.

Theorem. *The identity*

$$\sum_k \frac{(a+k-1)!(b+k-1)!(n-k-a-b+c-1)!}{k!(n-k)!(k+c-1)!} = \frac{(n+c-a-1)!(n+c-b-1)!}{n!(n+c-1)!}.$$

is true.

Proof: Take

$$R(n,k) = -\frac{(b+k-1)(a+k-1)}{(c-b+n)(c-a+n)}. \blacksquare$$

Now here is how to complete the proof of an identity if such a rational function be given. First, divide through the identity by its right hand side, so it takes the standard form $\sum_k F(n,k) = 1 \quad (n \geq 0)$. Next, define a function $G(n,k) = R(n,k)F(n,k-1)$, where R is the rational function that is given in the 'one-line' proof certificate. Then verify that the pair (F, G) satisfy the WZ condition

$$F(n+1,k) - F(n,k) = G(n,k+1) - G(n,k).$$

Finally check that the boundary conditions $G(n, \pm\infty) = 0$ are satisfied.

Having done that, you have proved that $\sum_k F(n,k) = const.$, for all $n \geq 0$, i.e., you have proved the identity.

Since the process of verifying the proof certificate is so mechanical, the whole job can be done by computer, and the task of proving complicated identities can be removed from the list of human tasks.

Such rational function certificates exist for all of the classical hypergeometric identities, such as those of Saalschütz, Dixon, Whipple, Dougall, Clausen, and for a host of newer hypergeometric identities as well. As is well known, these hypergeometric identities imply huge numbers of binomial coefficient identities, and so all of these have certificates too. For more details see [WZ].

References

[Eg] Egorychev, G.P., Integral representation and the computation of combinatorial sums, American Mathematical Society Translations, vol. 59, 1984.

[GKP] Graham, Ronald L., Knuth, Donald. E., and Patashnik, Oren, Concrete Mathematics, Addison-Wesley, 1989.

[Gou] Gould, Henry W., Combinatorial Identities, Morgantown, W. Va., 1972.

[Go] Gosper, R. William, Jr., Decision procedure for indefinite hypergeometric summation, Proc. Nat. Acad. Sci. U. S. A. **75** (1978), 40-42.

[Kn] Knuth, Donald E., The Art of Computer Programming, vol. 1: Fundamental Algorithms, 1968 (2nd ed. 1973); vol. 2: Seminumerical Algorithms, 1969 (2nd ed. 1981); vol. 3: Sorting and Searching, 1973, Addison-Wesley.

[Ro] Roy, Ranjan, Binomial identities and hypergeometric series, Amer. Math. Monthly **97** (1987), 36-46.

[WZ] Wilf, Herbert S., and Zeilberger, Doron, WZ pairs certify combinatorial identities, preprint.

Department of Mathematics
University of Pennsylvania
Philadelphia, PA 19104-6395

Printed in the United States
By Bookmasters